THIRD EDITION

CONCEPTS
IN
WINE
TECHNOLOGY

SMALL WINERY OPERATIONS

YAIR MARGALIT, PH.D.

THIRD EDITION

Concepts

in

Wine

Technology

Small Winery Operations

Yair Margalit, Ph.D.

The Wine Appreciation Guild
San Francisco

Concepts in Wine Technology
Small Winery Operations
Third Edition

A Wine Appreciation Guild Book
All rights reserved
Text copyright © 2004, 2012, 2018 by Yair Margalit, Ph.D.

Managing Editor: Bryan Imelli
Copy Editor: Judith Chien
Assistant Editor: Corrine Cheung
Text and Cover Design: Balzac Communications

Library of Congress Cataloging-in-Publication Data
Margalit, Yair, 1938-
Concepts in wine technology : small winery operations /
  Yair Margalit. – 3rd edition.
pages cm
Includes bibliographical references and index.
ISBN 978-1-935879-94-7
1. Wine and wine making. 2. Wine and wine making--Chemistry.
  3. Small business. I. Title.
TP548.M338 2012
663'.2--dc23
2012027550

Manufactured in the United States of America

The Wine Appreciation Guild
and imprint of
Board and Bench Publishing
www.boardandbench.com

CONTENTS

Introduction

After my book *Concepts in Wine Chemistry* was first published, I felt that another, more technically oriented, was needed, one which would better cover the practical aspects of winemaking. The new, third editions of *Concepts in Wine Chemistry*, and its sister text, *Concepts in Wine Technology* are together a comprehensive treatment of theoretical and practical winemaking. Both are based on my academic background as a physical chemist and lecturer in wine technology and wine chemistry and on my experience as winemaker and grapegrower at Margalit Winery.

In this new edition, we decided to modify the approach to the subject matter, due to the significant changes in the field since the publication of the previous edition.

The book is constructed to cover the entire process of wine production in small wineries, where the winemaker is intimately involved in all its aspects. It contains seven chapters starting with grape ripening and pre-harvest operations, through harvest, fermentation, cellar operations, barrel aging, bottling, and a final chapter that covers phenolic compounds, oxidation and micro-oxygenation, sulfur-dioxide, wine faults, legal regulations, and wine evaluation.

CHAPTER I

PRE-HARVEST

CHAPTER 1: PRE-HARVEST

A. Grape Ripening

The main object in following up the developments and changes of the grapes during the ripening period is to confirm the maturity of the grapes, with respect to the wine we are going to produce. Determining the maturity of grapes (or more specifically the time to harvest) is not a simple and clean-cut decision; it depends on many factors such as the variety of the grapes, the appellation (soil and climate), the weather during the last days or weeks before the prospective time of harvest, the bunch health (mold or insect infection), the vineyard disease state, and the style of wine to be made.

The development of the berries from the time of fruit set until ripeness is roughly sketched in the following figure. At the beginning of stage III the berries begin to change color and soften. Just after these changes, the development of the berries is very fast, followed by a rapid increase in the sugar concentration. Stage IV, where the grapes are over-ripe, is characterized by further softening of the berries' skin, shriveling due to water evaporation, and finally drying or mold infection. The exact time to harvest, as will be discussed in this chapter, is somewhere in the interval between the end of stage III and up to a certain point in stage IV.

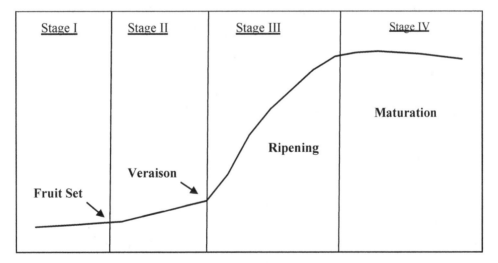

Schematic figure of grape development. The X-axis is the time. Y-axis can be any berry parameter, such as sugar concentration or berries' weight.

Fast changes occur during stage III, where frequent samplings are needed in order to follow the maturing changes of the grapes.

The chemical changes one should follow are: sugar concentration, acid content, and pH level. Start following these changes when the grapes' color begins to change. White grapes will turn from a deep green to yellow-green, and red grapes from green to purple-red. The change is quite rapid, a matter of a week or two. These changes usually happen at a sugar level of about (15 – 17) Brix.

1. Sugar

The main sugars in grapes are *D-glucose* and *L-fructose*, generally in 1:1 ratio, with fluctuations of ± 30% depending on the variety and maturity of the grapes. In grapes infected with *Botrytis-Cinerea* mold, the ratio is in favor of fructose, which is twice as sweet as glucose. During fermentation with most yeast strains, the consumption of glucose is faster than that of fructose, and toward the end of fermentation, most of the residual sugar is therefore fructose. Some other sugars which are found in grape juice are *sucrose* at about (2 – 10) gr/L, *L-rhamnose* (0.2 – 0.4 gr/L), *L-arabinose* (0.5 – 1.5 gr/L), and *pectin*, which is a higher molecular weight sugar at about (0.2 – 4.0) gr/L. The actual range of glucose and fructose concentrations in grape juice is (80 – 130) gr/L each.

Measuring Sugar Content in Grape Juice

Being the major soluble solid content in grape juice (about 95%), the measure of must density will reflect fairly well the total concentration of sugars in the must. From the tradition of the trade we have inherited three unites for sugar concentration, namely, *Baumé, Brix,* and *Oechsle*.

Baumé is approximately the potential alcohol %, which will be produced by fermenting the sugars to dryness. Its use in the wine industry reflects directly the potential alcohol content in % (volume/volume).

So for example, a must containing 12 Baumé will ferment to wine having about 12% alcohol (vol/vol).

Oechsle is the must density reading, translated to more workable numbers, namely, the density value minus 1.0 times 1000, or in a shorter presentation **Oechsle = (d – 1.0) x 1000** (where **d** is the must density). For example if the density of the must is 1.074 (which corresponds to 9.9 Baumé), its Oechsle value is (1.074 – 1.0) x 1000 = 74.

Baumé is still in use mainly in France, and Oechsle in Germany, Austria, and Switzerland.

Brix (B°) (also called *Balling*) is the % of total solids in solution, in grams of solute per 100 gram solution (gram/gram). This unit was very much used in general chemistry practice, and in recent years it is the most used unit for sugar content in the wine industry.

All the above units can be easily measured by two physical methods: density by hydrometer, or optically by a refractometer. In the first method, the hydrometer is placed in the must solution and the sugar concentration is read on its long stem floating above the liquid surface. The hydrometer reading can be used to measure the must's Baumé, Brix, or its density. The relationship between the density (at 20°C) and the major units in use, Baumé and Brix, is shown in the following table:

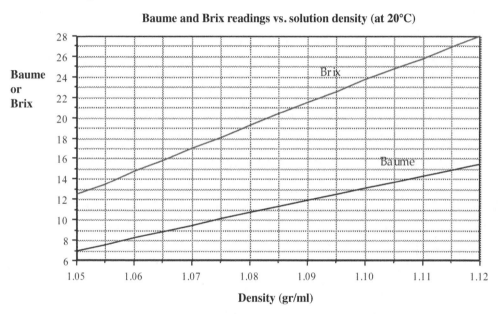

Baume and Brix readings vs. solution density (at 20°C)

In order to transform between Brix-Baumé units, one should take the actual number in the measured system (Brix or Baumé) in the graph and follow the vertical direction (Y-axis) to the other system to read its equivalent value. For example, 10 Baumé leads upwards to its equivalent value of 18 Brix, or 26 Brix leads downwards to 14.5 Baumé. The two examples above have density readings of 1.075 and 1.11 respectively.

The special hydrometers that measure the Brix (or Baumé) reading *directly* are calibrated at 20°C. For measuring must at a different temperature, a correction should be made. At temperatures above 20°C the correction value should be added to the hydrometer reading, and below 20°C the correction value should be subtracted. The following correction figure might be helpful:

Brix value corrections at different measuring temperatures

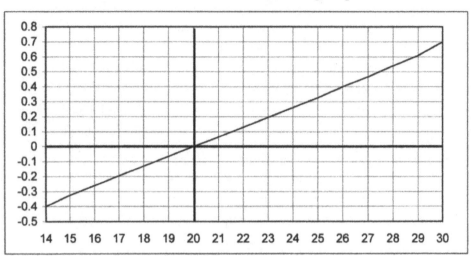

Measuring temperatures °C

For example, if you read 22.3B° at 26°C, then the correct Brix content is 22.7B°. Since the Brix (measured by density) includes sugar and all other solid materials in the must (such as acids, minerals, phenolic and pigment compounds, flavor compounds, proteins) the approximate content of pure sugar can be estimated to be (0.90 – 0.95) B°.

The sugar is converted by the fermentation, mostly to ethyl alcohol and carbon dioxide:

$$(1) \quad C_6H_{12}O_6 \text{ ---> } 2\,C_2H_5OH + 2\,CO_2$$
$$ \quad \text{180 gram} \qquad \text{92 gram} \qquad \text{88 gram}$$

This means that each sugar molecule is converted to two alcohol and two water molecules. And quantitatively, 180 gram of sugar produces 92 gram of alcohol and 88 gram of CO_2.

From the molar ratio of the products (alcohol and CO_2) to substrate (sugar), 51% of the sugar is transformed into ethyl alcohol (92/180 = 51). In practice, the yeast itself utilizes part of the sugar, and some part is converted to other products (higher alcohols, aldehydes, esters, acids, etc.). Therefore, the molar yield of ethyl alcohol production during fermentation is only around 47%.

Careful calculation can be made in order to have a good estimation of the potential alcohol content in the finished wine from the Brix measurement. The result (expressed in the common volume /volume unit) is expressed by:

1.75 B⁰ ===> 1% (v/v) alcohol. Or as:

(2) $\boxed{\textbf{\% alcohol (v/v) = 0.57 x Brix}}$

This expression is valid in ripe grapes within a Brix range of (18 – 25) B°, with accuracy of ± 0.2% of the potential alcohol. Thus, for example a must with 22.7B° will end with (12.9 ± 0.2) % alcohol. The final ethanol content is also slightly dependent on the fermentation temperature, due to evaporation losses.

As mentioned above, the second method for measuring the sugar concentration is the refractometer, which is based on the refractive index of sugar (linear dependence on concentration). This tool is handier than the hydrometer and therefore it is very useful for field tests. Its values are slightly lower than those measured by the hydrometer.

2. Acids

We shall start with some words about acidity, to emphasize its importance in grapes and wines. Acidity is one of the most influential factors in wine. It affects the wine's microbial stability, malo-lactic fermentation, color, and aging rate, as well as its tartrate and protein stability. But above all, acidity has a major effect on wine's tasting balance perception.

Acidity in wine has two sources: acids that develop in the grapes (mainly *tartaric, malic, citric*), and acids formed through the process of winemaking (*succinic, acetic, lactic* and others). We shall mention here some of the major acids in grapes and wine, but will start with the general aspects of acidity which are relevant to winemaking. Grape acids are considered to be weak organic acids, and the following section will deal with the behavior of such acids in aqueous solution. An acid is represented by AH, where the A symbolizes the basic structure of the acid and H is its dissociated proton which can be separated (dissociated) from the basic structure A. When dissociation occurs, the results are a positive charged proton H⁺ and negative charged acid A⁻. A charged atom or molecule is called an ion. The dissociation has three basic characteristics:

It is reversible, dynamic, and it reaches some equilibrium between the parties. The ionic dissociation of acid **AH** is represented by a general formulation:

$$AH \Longleftrightarrow A^- + H^+$$

Meaning that the acid AH in aqueous solution, is in equilibrium with its dissociated components, the positive hydrogen H^+ (proton) and the negative part A^- (the acid anion).

A *dissociation-constant* K_d is defined as: $\quad K_d = [A^-][H^+]/[AH]$

where $[A^-]$ and $[H^+]$ are the equilibrium concentration values (molar) in the solution of the anionic form (A^-) of the acid and its proton (H^+) respectively.

$[AH]$ is the un-dissociated acid concentration.

In other words, the result of multiplying the molecular concentration of the acid anion by the proton concentration and dividing by its undissociated acidic concentration is a constant value. This constant K_d expresses the equilibrium between the parties. Each such acid has its specific K_d.

The K_d expresses the *strength* of the organic weak acid, namely, the greater the K_d (which means the concentrations of $[A^-]$ and $[H^+]$ are higher), the stronger is the acid when compared with another one which has a smaller K_d. In other words, the strength of an acid is a measure of its ability to release $[H^+]$ into solution.

The following definitions will lead us to a very useful presentation of the relationship between acid concentration and one of the most important terms in winemaking, the pH:

Operating a logarithm on both sides of the K_d definition will lead to:

$$K_d = [A^-][H^+]/[AH]$$
$$\log K_d = \log [A^-]/[AH] + \log[H^+]$$

changing signs in both sides: $\quad -\log K_d = -\log [A^-]/[AH] - \log[H^+]$

and by new *definitions* : $\quad -\log K_d = pK_a \quad$ and $\quad -\log[H^+] = pH$
$$===> pK_a = -\log [A^-]/[AH] + pH$$

By rearranging the order in the last equation we get a very useful expression of the relationship between acid dissociation in solution and its pH value by knowing its constant number pK_a.

(3)
$$\boxed{\log [A^-]/[AH] = pH - pK_a}$$

This relationship is very basic in dealing with weak acids. It allows one to calculate the *dissociated* to *undissociated* acid ratio, namely, $[A^-]/[AH]$ at a given pH, or to calculate the pH of an acid solution by knowing this ratio.

In wine, the pH is a most important factor, as it reflects the actual proton concentration in solution, which is not necessarily the total acid concentration.

In strong inorganic acids where acids are practically 100% dissociated, K_d is practically infinitely big and the pH (in logarithmic scale) represents the *actual* acid concentration. But in weak acids, as is the case in wine, where dissociation is very partial, the above equation defines the connection between the actual acid concentrations and their pH value.

Most wine acids are di-acids (capable of releasing two protons per one molecule), so a second dissociation takes place with a second dissociation-constant that is always very small in relation to the first one:

$$AH_2 \overset{K'_d}{\Longleftrightarrow} AH^- + H^+ \overset{K''_d}{\Longleftrightarrow} A^= + 2H^+$$

or

$$AH_2 \Longleftrightarrow A^= + 2H^+$$

In grape juice and wine there are some organic weak acids, which together contribute to the wine acidity and its pH. Each one of the acids has its typical pK_a and its actual concentration range. Together with another major factor, namely, the concentration of the alkaline metal potassium, the wine pH is set up. The calculation of the expected pH in such solution containing weak acids and alkaline metal is beyond this chapter, but it is helpful to know the final result that the combinations of the total acids and potassium ion (K^+) in wine bring its pH range to span from **2.8 up to 4.0**. (If the reader is interested in the full calculation guide, please refer to the *Concepts in Wine Chemistry* book).

White wines usually have a lower pH (3.0 – 3.5), while red wines usually have higher pH values (3.3 – 3.9). Sparkling wines generally fall in the very low pH range (2.9 – 3.1) because grapes for sparkling wines are harvested early, before they reach full maturity, when their pH is still very low.

The major acids in wine are presented in the table at the top of the following page, with their molecular weight, dissociation-constants K_d, and their pK_a (at 25°C). It is important to distinguish between the terms *total-acidity* and *titratable-acidity* (TA). The total-acidity is the concentration of the *total organic acids* in wine. *The titratable-acidity is the total proton concentration in wine, which is measured by titration of the wine with a strong base*, and it is not equal to the total acidity. It is less. The alkaline ions which partially neutralize the organic acids account for the difference. The total acidity is therefore equal to the sum of the concentrations of titratable protons plus the potassium ions (which is equal to the proton ions that are neutralized by it).

Acid	Molecular	K_d		pK_a	
	Weight	First	Second	First	Second
Tartaric acid	150	9.1×10^{-4}	4.2×10^{-5}	3.04	4.34
Malic acid	134	3.5×10^{-4}	7.9×10^{-6}	3.46	5.10
Citric acid	192	7.4×10^{-4}	1.7×10^{-5}	3.13	7.47
Succsinic acid	118	6.1×10^{-5}	2.3×10^{-6}	4.21	5.64
Lactic acid *	90	1.4×10^{-4}	--	3.86	--
Acetic acid *	60	1.8×10^{-5}	--	4.75	--

* Mono acids (one acidic proton per molecule)

Therefore the *total acidity* $[A]_{total} = [H^+]_{titratable} + [K^+]$
where $[H^+]_{titratable}$ is the titratable acidity measured by titration, and $[K^+]$ is the potassium concentration. The combination of weak organic acids and alkaline metals in wine makes it a buffer to pH changes, meaning that wine has some resistance to pH changes by acids or base additions. The buffering capacity depends on the total acidity.

Grape Acids
Some details on specific grape acids are given here:
   Tartaric acid: COOH – CH(OH) – CH(OH) – COOH is the key acid in wine. It is quite unique in nature, being found solely in vine fruits. It is also unique as being symmetrical in its structure on one hand, and containing two un-symmetrical carbon atoms on the other hand. Asymmetric carbon means that each of the four bonds of the carbon is bonded to a different atom or group of atoms (marked with bold **C** in the formula).

$$
\begin{array}{ccc}
\text{OH} & & \text{OH} \\
| & & | \\
\text{COOH} - \mathbf{C} & --- & \mathbf{C} - \text{COOH} \\
| & & | \\
\text{H} & & \text{H}
\end{array}
$$

The asymmetry causes this molecule to exist in four stereoisomers (different structures). Two of them are mirror images of each other and are defined as optical isomers. The great wine chemist Pasteur was the first to discover this unique and very important chemical phenomenon in the mid-nineteenth century. Only one of these optical forms (out of the four possible isomers) is found in grape juice. Its notation is (L+) tartaric acid. (For more details the reader is referred to *Concepts in Wine Chemistry* book.) Its concentration range in grapes is about $(1 - 7)$ g/L.

No chemical changes occur to this acid during vinification except partial precipitation as potassium-bitartrate.

Malic acid: $COOH - CH(OH) - CH_2 - COOH$ is also a major acid in grape juice, but it is a very common acid in many other fruits. It has only one un-symmetric carbon (bold **C**) and hence only two isomers. The natural acid in grapes and other fruits is L(-) malic acid isomer. Its concentration range in grapes is about $(1 - 4)$ g/L, depending on the maturity state.

Before ripening the two acids are at about the same concentration. During the advanced maturing process, toward the end of stage III, the concentration of malic acid gradually and consistently decreases by respiration, while that of tartaric acid remains practically unchanged. At full maturity the concentration of tartaric acid is always higher, where the tartaric/malic ratio is found mostly in the range of $(1.1 - 2.0)$, depending on the variety, climate, and maturity state. As a result of the malic acid respiration during maturation, the total acidity is reduced, and at the same time the concentration of potassium is increased. These two processes cause the grape juice pH to increase, mainly at the end of the ripening process (stage III – IV). Higher pH means lower acidity.

Malic acid is prone to chemical change by malo-lactic bacteria (to be discussed in detail in chapter III, section F), so its concentration may decrease very much before the wine is finished.

L-citric acid: is another acid which exists in must in a low concentration, with a range between $(0.1 - 0.3)$ gr/L. **It rarely exceeds 0.7 mg/L.**

The total titratable acidity of all acids in must is in the range of $(4 - 12)$ gr/L. Among these acids, tartaric acid, having the highest K_d (and smallest pK_a) among wine's acids, is the strongest acid in must and wine, and it contributes the most to its pH and acidity (see table above). Notice also that the two major acids, tartaric and malic, are *di-acidic*, which means that the second proton also contributes to their total acidity but not to the pH, because the second dissociation constant is far lower (4.25×10^{-5} and 7.9×10^{-6} respectively) than the first one.

3. Maturity

There are some different criteria for determining the maturity stage of the grapes and therefore the optimal time for harvesting. It depends very much on the grape variety, the type of wine being produced, and the climate and microclimate of the vineyard. It is also very dependent on the annual climate at each specific vintage. Therefore, harvest time might change drastically every year at the same vineyard. In any method being used to determine maturity, there is practically no need to start to collect grape samples before the grapes have reached (16 – 17) Brix. The profiles of the major parameters during the weeks of final maturation have the following shapes:

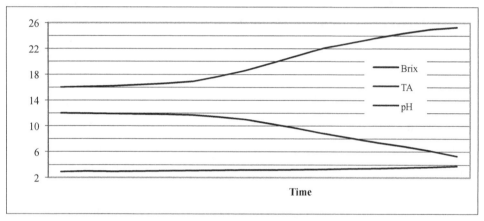

Brix, TA, and pH profiles during the final states of maturation (stages III and IV. The Y-axis represents the actual range values of these parameters.

By measuring the changes in the sugar, TA, and pH over certain time intervals, one can follow the grapes' development towards maturity. Yet the main question is still unanswered: what criterion is the most important guidance to the time of harvest?

For example, if the sugar is allowed to develop to very high level, other parameters could become undesirable. The TA might be too low, the pH too high, and/or the aroma materials might change their nose to a "jammy" or baked flavor. Also, waiting too long might expose the grapes in certain areas to weather risks (rain, high temperatures), or insect and mold infections. There must be a certain point during maturation which is the optimal one, when all the considerations are best taken into account. When is that point?

We shall skip all other methods found in the literature and bring here the one that is, in our experience, the most valuable and practical. It considers mainly the sugar development. As long as sugar is still accumulating (which of course is not always the case), all other parameters are more or less moving in the right direction. At each sampling, calculate the Brix/day accumulation (since last sampling). When it comes close to 0.1 Brix/day, start to think about harvesting. If it reaches 0.05 Brix/day, the maturation is close to being stuck and no further waiting is advised. This method is simple and, in most cases, satisfactory.

In cold growing regions where grape ripening is a well-known problem (e.g., Germany, certain places in France, and other northern countries), the major wine quality parameter is the sugar content of the grapes at harvest. Also in other places during certain years, the weather conditions (such as cool summer, cloudy skies) may sometimes cause a long delay in ripening. If the delay is accompanied by rain and high humidity, the grapes will probably be infected with mold, which causes more problems as the harvest is delayed. Under such conditions there is no alternative but to harvest immediately, although the sugar and flavor have not fully developed. The deficiency in sugar level and high acidity may be adjusted, but the aroma and flavor of the wine may suffer. Unfortunately, such wine is likely to be of low-to-medium quality. In these cases the sugar content should be watched very carefully in the hope that it will reach the desired concentration. Even in moderate and warm countries, it may happen that the maturation of the grapes is halted for unknown reasons. The Brix does not go up, the TA is continuously reduced, and the pH begins to run sky high. Waiting in such situations, with the hope of developing more sugar, is probably a mistake. The quality of the grapes will likely decline.

In view of the above-mentioned risks, it is very important to watch the smooth development of the Brix until it shows signs of slowing down. This is vital for finding the right time to harvest. But a very important point should be remembered: there must be a limit of how high to let the Brix to develop when there is no sign of halting its growing. The high sugar level will produce alcohol which might be too high for table wine. Above 14% it starts to feel hot in the mouth and unbalanced (a flaw which is quite common in certain wines). So a limit of about (24.5 – 25.5) B^o should be the maximum.

There are some general guidelines to help us to relate to the different parameters according to grape variety. These parameters depend very much on the terroir where these varieties are grown, and should be studied in each case. As a very general guide, we can use the following recommended *range* for the main parameters Brix, TA, and pH for white and red grapes:

Type of wine	Brix	TA (g/L)	pH
White	20 - 23	6 - 9	3.1 - 3.4
Red	21 - 25	5 - 7	3.3 - 3.8

If in white grapes, for example, the Brix is still low (around 18 Brix), the TA has dropped down to below 6 g/L, and the pH is above 3.4, one should probably wait for the Brix to rise, and make the necessary corrections for the deficiency in acid and high pH. On the other hand, with the same TA and pH, if the Brix were close to (21 – 22), it would be better to harvest very soon.

In the case of red wine, if for example the Brix has reached 24.5B°, no matter what the other parameters are, it is better to harvest. And if, say, it is 23.3B° and the pH went up to 3.8, it would be better not to wait any longer, although acid correction is an option.

Another very important aspect to consider when following grape maturity is the varietal aroma and flavor. With this in mind, the Brix level need only be kept within the general recommended range. The Brix limits are important only for maintaining the alcohol range in the finished wine. The full varietal flavor in the grape is most important for achieving the maximum varietal character of the finished wine. To the best of our knowledge there is not a practical measure to follow it in the grapes. So the only way to use it requires experience and the ability to project the vague perception of "grape flavor" and "taste" into the finished wine. Speaking from personal experience, this quality projection can be gained during many years of experience in a specific vineyard or appellation. There are some more complicated methods which reflect the ripening development, such as measuring the tartaric/malic acids ratio, the polyphenol compounds development, or the potassium development, but none of these methods includes a conclusive, precise assessment of the maturity timing.

4. Sampling

The technique of sampling is very crucial in order to get a real representation of the grapes' degree of ripeness. There are many reasons why right sampling is necessary. First, the distribution of the major components in the berries along the cluster is not even. The top berries in the cluster are higher in sugar than the bottom ones. After veraison a distribution of color can be seen in different locations of the berries in the cluster. In many cases even at full red color, some berries are still green. Second, the clusters themselves are not evenly developed, depending on their location on the vine, on the amount of light they were exposed to, and the location of the vine in the vineyard. On top of this, the content inside the individual berry is not uniformly distributed. Usually the sugar concentration is higher closer to the skin and drops down toward the center of the berry. The acid concentration follows the opposite trend. All the above factors are important to recognize, in order to get the most representative prediction of must sample from the vineyard. Based on all these factors, it is clear that individual berry sampling will not give any relevant data. We need to collect whole clusters to represent the average grape maturation stage. Also the clusters have to be collected in such a way as to represent the whole lot of the vineyard, and it is better to do it on not too big a lot.

The sampling technique which might give quite a good picture of the grapes' maturity is to collect about (50 – 100) clusters from different locations on the vine (both sides of the row), lighted and shadowed, and from different vines in the area. It is advisable not to collect the sample from a very large area of the vineyard (not more than 5 acres per sample), because of the wide variability due to microclimate and soil in the same vineyard. And if the topographical variability of the vineyard is high (slope, different directions to the sun, valley, different soil), the sampling should be divided to smaller lots according to the specific sites.

The simplest way to pick the clusters is to walk in the vineyard along the rows, and for every certain number of vines, to pick a cluster from both sides of the row. The sample should be weighed and the number of clusters be counted in order to establish the average weight of a cluster. Counting the average number of clusters on each vine may lead to an estimation of the yield in that vintage. The sample has to be crushed by hand as firmly as possible, and the juice separated from the skins and seeds by a simple net. The Brix, total acidity and pH should be measured and be placed on a graph. The Brix/day change since the last sampling should be marked. The sampling and measurements should be taken once or twice a week, depending on how close the grapes are to finishing their maturation process.

5. Composition and Yield

The overall weight composition (%) of a grape cluster is:

Cluster	Range (%)	Average (%)
stems	2 - 6	4
seeds	0 - 5	4
skins	20 - 30	25
must	60 - 70	67

The lower figure in the must range (60%) is for the free-run juice, and the upper one (70%) includes the fully pressed juice (or wine).

As for the clear must, its composition is:

Must	Range (%)	Average (%)
water	75 - 85	77
sugars	17 - 25	22
organic acids	0.4 - 1.0	0.6
other dry extract	0.3 - 1.0	0.5

The other-dry-extract contains proteins, amino acids, alcohols, polyphenols, esters, minerals, and aroma components.

From the wine production perspective the yields are:

White Wine: About 60% of the grapes' weight (made out of free-run only) and close to 70% of the grapes weight (when press-run included) ends up as white wine.

Red Wine: About 60% of the grapes' weight (free-run) and up to 65% (press-run included) ends up as red wine. (The small difference in yield between white and red wines comes from the difference in the skins of white and red grapes. The former have in general thinner skins and bigger berries.)

These figures of production yield can be expressed in a different way: from every ton of grapes (1000 Kg) red or white, one should get about 60 to 65 cases of free-run wine bottled in 750 ml bottles, at 12 bottles per case. If press-run is included, another 3 – 7 more cases can be expected.

B. Pre-Harvest Operations

The winery's policy for the coming vintage should be decided before planning the major aspects of winery operations. This should include the following points:

- Types and qualities of wines to be produced.
- Quantities of each type.
- Grape sources (quantities, quality and prices if bought from other growers).
- Financial and budget plans.
- Marketing prospects and strategies.
- Availability of manpower for the crush season.

1. Vineyard Management

From veraison up through harvest, careful vineyard management has a close relationship with wine quality and production. For best results, the winemaker should pay close attention to:

Chemicals used against different kinds of insects and fungus: The chemicals are mainly used as preventive agents during shoot and leaf growth (March-July in the northern hemisphere) and are needed much less toward the end of the summer. At this time the main concern is the bunch rot caused by different insects, and microbial agents such as *Botrytis Cinerea*, *Aspergillus*, *Penicillium*, *Acetobacter* and others.

Some chemicals may have residual activity causing health hazards, and some (especially those preventing fungus infections) may inhibit fermentation. For example, when sulfur dust is used (against powdery mildew and other fungus agents), it may be carried into the wine by the grapes, and may cause production of hydrogen-sulfide during fermentation, which is an unpleasant odor compound. "Bordeaux Soup" (copper sulfate, widely used as a repellent against downy mildew in Europe) will carry copper into the must and may later cause copper haze in white wine (if the concentration of copper is above 0.5 ml/L). It may also inhibit yeast growth during fermentation.

More efficient chemicals are available now, which can focus on their specific targets with minimal environmental or health effect.

In order to prevent any chemical from being carried onto the grapes and into the wine, their use should be terminated sometime after veraison (color change of the grapes). The time remaining before harvest should be long enough for the chemical agents to decompose, evaporate, or be swept away by wind or rain.

17

Irrigation during the summertime when there is no rain: If irrigation is continued up to harvest, it may dilute the dry extract of the berries and reduce grape quality. It is believed (with no firm evidence) that the vine should struggle and be under some stress during the last period of ripening in order to produce a good quality wine. It may be true or not, but in any case, extra water at this stage will be used by the vine mostly for shoot growth or delivered to the berries, diluting their content. Therefore, it is generally recommended to stop irrigation at least a month before the harvest. Care should be taken in hot summers or in hot regions, because drying the vines too early may cause the leaves to turn yellow and stop photosynthesis. Consequently, the maturity could be delayed or even worse, not come at all. Therefore, very careful inspection should be carried out with regard to irrigation management. Probably the best method is to irrigate small quantities once or twice a week, just enough to keep the vine from turning yellow. Also, it might be a good idea to cause the vine to dig its roots way down into the ground, reaching for last winter's rainfall water instead of irrigation water. This can be done over a few years (better when the vines are still young) by managing long intervals between each irrigation, while using large quantities of water each time.

Leaf removal and pruning: After veraison, it is important to prevent the clusters from being over-shadowed. Berries exposed to light develop better flavor and higher sugar levels than shaded ones. One efficient way to improve light exposure is to remove the leaves over grape clusters. It's a labor-intensive operation, but it leads to higher grape quality. Leaf removal should start after veraison, and should be repeated whenever necessary.

Another factor to consider is pruning. Long shoots on the vine do not contribute to the berries' development. They also create obstacles for hand harvesting as well as mechanical harvesting. Thus, it is good practice to prune long shoots, clearing the way in the vineyard and decreasing shadow areas on the vine. It is recommended that this pruning be done about (2 – 4) weeks before harvest, depending on the vigor of the vines.

Harvest: The traditional harvest is done by hand and, for small wineries, this is much preferred over mechanical harvesting. Mechanical harvesting is used mainly by large wineries because of the low availability of manpower for hand harvesting, which is hard seasonal work with low pay. In addition, mechanical harvesting allows the grapes to be picked at night when temperatures are low, which is a big advantage in warm regions. However, the main disadvantage of mechanical harvesting is that most of the berries' skins are broken, releasing juice into the container. This can lead to oxidation and unwanted fermentation. Yet the most threatening factor of a mechanical harvest is the time allowed to pass between the harvest and the processing of the grapes in the winery. The shorter the time, the lower is the risk of damage. Another disadvantage is the quite high content of leaves collected into the must by the harvest machine.

On the other hand, with hand harvesting, damage to the berries is minimal, so oxidation and uncontrolled fermentation are almost entirely prevented. Yet some precautions should still be taken. For example, when hand harvesting, it is recommended to start picking very early in the morning until noon, so the grapes can be picked at cooler temperatures.

2. Preparing the Winery for Harvest

The mid-summertime before the harvest is the best time to prepare the winery for its new vintage. Everything needed for the harvest should be made ready. This is vital because during harvest there is no time to look for missing chemicals, fix broken machinery, patch leaking valves in a tank, etc.

Tanks: The estimated quantities of grapes of each variety, determines the volume needed in individual tanks and the total tank volume needed for the coming vintage. During fermentation, white varieties need about 95% of the tank volume, while red fermenting varieties need about (80 – 85) % of the tank volume to allow some volume for the cap.

After fermentation and skin separation of red grapes, the must volume reduces to about (60 – 70) % of the original volume. So the volume needed during fermentation is greater than the storage volume by some (20 – 30) %. Each fermenting period is short, but the harvest season is quite long due to the various varieties' maturation dates. So careful calculation is needed, based on past experience, to plan the various tanks available and the planned crushing scheme. Matching expected yield of each variety and/or vineyard lots with tank volume is important. New barrels may be needed and extensive bottling may be required to ensure enough empty tanks for the new coming wine. Each tank should be checked for corrosion, valve defects, and any leakage problems.

If some barrel fermentation is planned for white wines, the new barrels should be ready and checked for possible leakage. The number of barrels should be carefully matched to the quantity of wine planned to be fermented in them. Each 225L barrel will contain the must of about (350 – 400) Kg of grapes with 2% of overhead space needed during fermentation. A clean, cool space should be reserved for this purpose.

Cooling system: The cooling unit should be in good mechanical condition, including the cooling liquid (usually ethylene glycol), which has to be checked for proper operation in the whole system. All pipelines from the cooling units to the tanks and heat exchangers should be leak-free and thermally isolated. It is highly recommended to test the whole system for its total cooling capability, and every individual tank for its thermostatic control operation. Detailed inspection of the cooling system might save a lot of problems later

Machinery: All processing machines, such as scales, hoppers, conveyors, crushers/destemmers, presses, pumps, lees filters, hot water pressure machines, and forklifts should be in good working condition. The machines should be checked for any potential mechanical or electrical problems, and appropriate measures should be taken. The hoses should be checked for leaks, and enough hoses of different sizes and lengths should be ready for use.

Chemicals: All chemicals used in the winemaking process should be in the winery, each one according to the amount needed with some to spare. The chemicals to be prepared are:
- Sulfur dioxide (as powdered potassium-meta-bisulfite, or as liquid in steel tanks)
- Tartaric acids
- Yeast in various strains needed
- Malo-lactic bacteria (frizzed dried powder)
- Yeast nutrient
- Di-ammonium-phosphate
- Fining agents (bentonite and any other needed)
- Filter pads of needed sizes, and diatomaceous earth if in use)
- Cleaning and sanitary materials

Laboratory equipment and chemicals: The laboratory should be ready for must and wine analysis. All the equipment for the basic analysis of Brix, titratable-acidity, pH, volatile-acidity, sulfur dioxide, malo-lactic fermentation, alcohol, and residual sugar should be ready, including calibrated solutions. All other chemicals needed for the general laboratory work should be on hand.

Sanitation: A clean winery may prevent some microbiological troubles during the wine processing. Any piece of equipment that will be in contact with the must or wine (machines, tanks, hoses, barrels) should be cleaned with water or with suitable cleansing agents, and rinsed with plenty of water. The whole floor area of the winery should be cleaned with a hot-water pressure machine. (This machine is the most important one in the winery!)

The sewage system of the winery must be checked for possible blockage or leakage. If there is any risk of water shortage during harvest time, it is highly recommended to have a full tank of water on reserve. A winery with a sudden water shortage faces a very serious problem.

Chapter II

Harvest

CHAPTER II: HARVEST

A. Destemming and Crushing

The purpose of destemming is to separate the stems from the berries, as they contain very high levels of tannins, and may contribute a hard "vegetal" or "green" flavor to the wine if the stems are fermented with the must.

The purpose of crushing is to break the berries' skins in order to release the juice. These two operations of destemming and crushing are generally done consecutively by one machine. In some machines, destemming is done first, and then the crushing. In other machines the order is reversed. We prefer the destemmimg first because less stem material enters into the must.

1. White Grapes

Fermentation of white must is carried out without the skins (unlike the red). Therefore, separation of the stems and the skins from the must before fermentation is a necessity. The destemmer/crusher, followed by press, is not the only method to do this. Direct pressing the whole clusters without destemming and crushing can also be used. The idea behind pressing the whole clusters is to minimize skin contact within the must. In this method the must that comes out of the press is very clear and can be ready for inoculation without further racking.

During mechanical treatment of the must, absorption of oxygen is very intensive due to the large surface area that the must is exposed to. Oxidation (followed by browning) of white must is very fast, especially if the must temperature is high (above 20°C), and is considered detrimental to the quality of the wine. The time taken for mechanical processing of the grapes, especially whites, should therefore be minimal. To protect the grapes, sulfur dioxide is added (see chapter VII, section C) as soon as possible. Some wineries do not add sulfur dioxide at this stage (in white wine), in order to oxidize the phenolic compounds on purpose, which polymerize, and precipitate during fermentation and fining. By doing so, the later browning of the white wine may be eliminated. This early (intended) oxidation of the white must, in contrast to the attempts to prevent it, is somewhat controversial. (For more details on this matter see chapter VII, section B.) After fermentation, during the rest of the vinification operations, the exposure of the wine to air by racking or filtering causes additional oxidation, which might add its share to the accumulating damage to the wine's quality. In certain cases where skin contact is desired, the crushed must can be transferred through heat exchangers in order to lower the temperature to about 10°C, and then to a drainer or a storage tank, for few hours, before pressing. If skin contact is not desired (as in most cases) the must can be transferred to the press, and through a heat exchanger to a settling tank.

2. Red Grapes

After destemming/crushing, the must is transferred to the fermentation tank to begin fermentation. Separation of the skins from the must will be done later, during or after fermentation, according to the winemaker's decision.

If "blush" wine is planned (to make white wine out of red grapes, *"blanc de noir"*), the must should be transferred from the destemmer/crusher to the press, and then through the heat exchanger to the tank for settling and racking. The must is treated exactly like white wine. By such fast skin separation, the color would be very light pink. If more color is desired, the must should be transferred to a tank for a few hours until the expected color is seen, and then transferred to the press.

The red skins of the blush juice coming out of the press contain the full concentrations of red grape components and can be added to the must of some other red wine (being in fermentation stage), to enhance its color, tannin, and body, if desired. This practice is called "bleeding" (*saignée*), and is sometimes used in the industry. In certain cases even the stems are used to enhance the tannin level to add complexity to the wine. This method is practiced in Bourgogne with Pinot Noir grapes, which are known to lack some color and tannin in their skin. The must containing the stems is then blended later with the rest of the wine.

B. Skin Contact

Skin contact is one of the most important aspects in the winemaking process. It influences the type, character, aging period, and general quality of the wine. Some of the varietal flavor and all of the color and tannin compounds in the berries are stored in the skin. (For details on the chemical structure of the phenolic compounds, and their functions in wine, see the chapter VII, section A.) These compounds are called phenolic. Roughly, they may be divided into three functional groups, responsible for *color* (pigments), *astringency* (tannins), and *flavor* (bitterness). The pigments contain mostly *anthocyanins* (a series of a few compounds such as cyanidin, peonidin, malvidin, petunidin, and delphinidin), which give the wine its red color. The color is pH-dependent and can be changed from deep red at low pH to green-blue at about pH 5 and above, with a gradual color transformation.

Tannins are polymeric forms of the phenolic compounds such as catechin, leuco-cyanidin, gallic-acid, vanillic-acid, and many others. Most polymers are composed of two and more monomers (up to about ten in wine), containing many OH sites, which are responsible for the "dry" sensation or astringency of red wine. The third group, the flavor compounds, is mainly derived from cinnamic acid derivatives (caffeic acid, coumaric acid, ferulic acid), and partly from the flavanols group. The above functional grouping is not strictly diverse, and some compounds may be colored and tannic, or tannic and flavored at the same time. The intensity and dispersion of these compounds in the skin depends upon many factors, such as variety, maturity state, climate, soil, and cultivars management. This chapter deals with the extraction of these compounds from the skins into the must.

During the vinification process, the extraction rate of the phenolic compounds depends on the variety of the grapes, as well as the temperature, alcohol concentration, sulfur dioxide concentration, and time of contact. The extraction is directly related to the last four factors. It is faster at higher temperatures, at higher alcohol and sulfur dioxide concentrations, and its concentration increases with length of contact (maceration). The total amount found in the finished wine is about $(1 - 2)$ gr/L in young red wine, and $(0.1 - 0.4)$ gr/L in white. In going into details about skin contact, we will separate the discussion on white and red grapes.

1. White Grapes

In white grapes, the skin contact intensity can be categorized into two groups: *no contact*, and *contact* which can be either *short contact* or *long contact*.

In *no-contact* treatment, the must is pressed immediately after destemming and crushing. The most extreme practice of this minimal skin contact is achieved by pressing the whole clusters without crushing. This technique is used mostly in sparkling wine production, where neutral basic wine with minimal character is desired, and also in grape varieties that have some bitterness in the skin, such as Muscat-related varieties.

Short skin-contact lasts between 1 and 4 hours, whereas *long skin-contact* can last for up to 24 hours. In these cases, the must is left with the skins after crushing, in order to extract more varietal aroma, knowing that there is also a gain of some undesirable bitterness from the flavonoid compounds, which later on might be oxidized and will turn the white color to yellow-brown.

When *skin contact* is desired, the must should be sulfited right after crushing, mixed well, and be cooled to (10 – 15) °C. The major effect of temperature on skin contact is the increasing rate of phenolic compounds extraction with increasing temperature. On the other hand, lowering the temperature at this stage is necessary in order to reduce the oxidation potential of the must. The length of skin contact depends on the variety, the ripeness of the grapes, and the kind of wine to be made. Some winemakers prefer minimal skin contact to get fresher, light white wine, while others prefer higher varietal character, with some risk of greater than normal tannin levels.

For aging white wine, it might be better to have some skin contact, which through aging the phenolic compounds will contribute to the bouquet of the wine. However, when the grapes are fully ripened, the length of skin contact can be shorter. On the other hand, if the grapes are not ripe enough and skin contact is long, care should be taken, because the must and later the wine may become "green" or get a "leafy" flavor, which may be undesirable.

After skin contact followed by pressing, the must contains a high content of lees that should be separated. This is done either by gravitational settling for 12 to 24 hours, or by centrifugation. In the gravitational settling, the lees settle down in the tank to form a dense layer. The clear must over the lees is then racked off to start the fermentation.

The lees left over can be filtered by a lees filter and be added to the rest of the must. (Good wine will not come out of the filtered lees.) Centrifugation, on the other hand, can reduce the solid content of the must at a very fast rate. The must's flow through the centrifuge is continuous, with interruptions from time to time to discharge the solids. The centrifuge can also be used at any other stage of processing, primarily after fermentation, to clear off the wine from the yeast lees. When centrifuging wine, care should be taken to prevent air from coming into the centrifuge (by using a nitrogen blanket), because this process introduces a lot of air into the liquid.

A somewhat different technique to treat white must is to use a drainer (juice separator). The drainer is a vertical tank that contains an inner screen (central or on its side), which the clear juice can flow through, leaving the skins behind. The drainer is filled with must directly from the crusher or through the heat exchanger, and the must is left in for a certain period, according to the desired level of skin contact. Then the juice is separated from the skins through the screen, either by gravitational pressure of the must, or by very light CO_2 pressure exerted from a CO_2 tank (up to 5 p.s.i.). The principle of the drainer is that the pomace itself functions as a "filtering" medium, reducing the solid particle content of the free-run juice. After the free-run juice has flowed out to another tank (by gravitation), the rest of the must is discharged into a press to extract the remaining juice. The pressed juice can be combined with the drained juice in a settling tank, where the total solids content is very much lower compared with a must that has not been drained. The whole process of draining (loading, waiting, draining the free-run, and discharging the must into the press) may take from about one hour, when minimal skin contact is desired, up to many hours as needed in a long skin contact.

Usually two or more drainers can serve one press. The drainers also function as a storage buffer between the destemmer/crusher and the press. The drained juice contains about (0.5 – 2) % solids, compared to (3 – 5) % in the usual press separation, which needs longer settling time to reduce this high solids content.

(Note: the Muscat varieties, such as Muscat of Alexandria, Muscat- Canelli, Sylvaner, Symphony, Emerald-Riesling and others, may develop a typical bitterness in their wines. For these varieties minimal skin contact may reduce this problem. Pressing the whole clusters without crushing may be the best method for such cases. On the other hand, in Chardonnay, skin contact is generally practiced, and in other varieties it depends on the case and on the wine's style).

2. Red Grapes

The phenolic compounds of the various phenolic groups in grapes are concentrated mainly in the berries' skins (about 50% of the total), in the seeds (about

45%), and the rest (5%) in the juice. The various groups are distributed as follows: Anthocyanins are located in the skins. The monomeric phenolic of the non-flavonoid groups and their polymeric tannins are found in the skins and in the seeds. And the cinnamic-acid phenolic group plus its esters are mainly in the juice and slightly in the skins. White grapes contain mostly these phenolic compounds. Therefore, there is a large difference in phenolic concentration in white and red wines. White wines contain a low concentration of phenolic, because it is made from juice that is pressed before fermentation. So only a small fraction of the phenolic found in the berries is introduced into the must.

In red wine production, the fermenting must contains skins and seeds during and after fermentation, which are extracted into the must/wine at a certain rate, and eventually, if enough time is allowed before pressing (separating the skins and the seeds), they may reach a concentration equilibrium between the skin/seeds cells and the outside media. The extraction is a diffusion process, which obeys basic diffusion rules, namely:

(1) The direction and rate of diffusion is controlled by the concentration gradient of the components under study. This means in simple words that the components flow from a higher concentration to a lower one and the rate of flow is faster when the concentration difference (the gradient) is higher.

(2) The diffusion rate is controlled also by the temperature. It is higher at elevated temperatures.

(3) If enough time is allowed, it will eventually come to a dynamic equilibrium between the two phases. The two phases are the skins/seeds cells and the wine. Final equilibrium will be reached only when the multiplicity of two factors, namely, the *concentration x affinity* in both phases, becomes equal.

In most cases the affinity between compounds to one phase, is not the same as to a second phase. Examples of two such phases are water and organic solvent, where mineral salt has very high affinity to water and almost zero to organic solvent. Water and membrane tissue of a living cell is another example. Therefore, one should keep in mind that only a certain percent of the color components and tannin contained in the grape tissue would diffuse into the processed wine, actually less than 50%. The rest of the phenolic will remain in the solid must particles (skin, seeds, pulp, tartrate crystals, and yeast cells).

A typical extraction rate of the color and tannin from the skins can be seen in thefigure at the top of the opposite page.

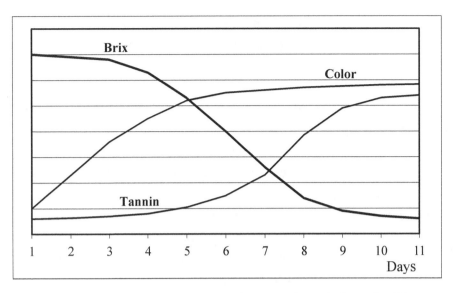

**Schematic extraction diagram of color and tannin from the skins
during and after fermentation.**

Therefore, separation of the skins from the must can be done at any time along the extraction line, depending on the grape variety and the wine style. If the separation is done shortly after the beginning of fermentation, the result is a good red color with a minimal amount of tannins, and the wine may be consumed very young. If separation of the skins is made after a long period of contact (long after the fermentation has been finished), the tannin extraction will be high. The wine is likely to be very tannic, hard, and will require longer aging. We can point out three kinds of skin contact treatment in red grapes, which will reflect the type of wine being made:

a. No skin contact: immediate separation of the skin from the juice after crushing. This wine, which is almost white or slightly pink in color, is called "*blanc de noir*" or "blush" wine. It is considered white wine, and all its processing is carried out accordingly.

Blush wine, which became very popular during the 1980s in the United States, is light, fresh, fruity wine, usually off-dry, with about (2 – 3) % residual sugar. The main problem with blush wine is that the bright pink color ages very quickly (in certain cases in a few months) to become brown-pink, which is not so appealing to the customers. It also loses its freshness and fruity aroma quite fast. The varieties that suffer the least from this problem are Cabernet Sauvignon and Zinfandel. A possible solution to minimize the color change is to avoid the use of SO_2 at crushing before fermentation, allowing part of the pigments to oxidize and polymerize, which then settle down later during and after fermentation. At bottling, with enough added free SO_2, the color of that wine will be more stable.

b. Short skin contact: before fermentation. The contact may last from a few to 24 hours. This "blush" wine is also called rosé, and the intensity of its pink color depends on the skin-contact length. The longer the contact, the more intense is the color. After press, to separate the skins, the must is processed as white wine to become rosé wine, which can be released in two months after harvest, and is best drunk within a few months afterwards.

c. Long skin contact: during and after fermentation. This is the classic red wine production, where the skin contact is going on during fermentation, and its duration may last from a minimum of 3 to 4 days for light red wine, up to the common practice of 2 – 3 weeks, and in certain cases even for a month or more in heavy, tannic, long-aging wines. The variety, the specific grapes, the style of the wine, and the expected aging time are factors that determine the length of the skin contact during and after fermentation.

Some grape varieties are intense in color and tannins (e.g., Carignane, Syrah, Cabernet Sauvignon, Zinfandel, Petite Sirah, Petit Verdot and others), while other varieties (especially Pinot Noir) have lighter pigments and lower tannin concentration. The color intensity also varies from season to season, and the decision when to separate the skins has to be made in accordance with each wine. The tannins are colorless, but they may contribute to the color intensity by co-polymerization with the red-colored anthocyanins to produce co-polymers (anthocyanin-tannin), which have a more intensely red color than monomeric anthocyanin itself. So, the desired tannin level in wine (responsible for the wine's astringency and bitterness) should be considered not only in prospects for aging, but also in its color intensity effect. For example, in Pinot Noir wines, which have low tannin levels, some wineries ferment certain portions of the must with the green stems (which contain a high tannin level), and blend this portion with the rest. The blend is higher in tannins and somewhat more complex (and "green"), but has a better chance for higher color intensity.

In a study measuring the tannin concentration in a very large sample of wines (1300 samples) made of many varieties and regions in the world, it was found that there is a big difference in phenolic concentrations between varieties on one hand, and a huge standard deviation within each variety up to half of the mean value. Cabernet Sauvignon, Merlot and Syrah are at the upper end, while Pinot Noir is at the lower end. In Cabernet Sauvignon, for example, the average tannin concentration was found to be (672 +/- 303) mg/L, and in Pinot Noir wines (350 +/- 200) mg/L as CE (Catechin Equivalent). The variability among grapes in each variety is so large that no conclusion can be made about the variety of the wine, based on its tannin concentration.

The factors that determine the final composition of phenolic in wine are:

Grape variety **and** *viticulture parameters*. The different varieties have typically different phenolic content, which finds its expression in the color intensity and tannin content of their wines. Also, "nature," namely, the location of the vineyard (climate, soil, elevation) and vintage are very important factors in this matter. The grape-growing practice, which includes all the planting parameters, such as irrigation policy, fertilization, canopy management and others, will also dominate the phenolic concentrations in a given vine.

Temperature of fermentation. Temperature profile during fermentation will affect the extraction rate. As the fermentation temperature becomes higher, the extraction rate gets higher. There are three reasons for this: one, increasing the permeability of the skins cell membrane at higher temperatures allows faster release of anthocyanin and tannin molecules outside; second, because of the solubility increase of those phenolic compounds in the wine solution at higher temperatures; and third, because of the diffusion of soluble compounds, which is faster at higher temperatures. This issue was investigated in many studies and was fully confirmed.

Maceration profile. At the beginning, when the must is kept in the tank before yeast addition, the diffusion occurs solely in a water solution where the color components (anthocyanin) are best soluble, causing their extraction rate to be faster. It slows down later when the alcohol content gets higher along with the fermentation. Also the concentration gradient becomes smaller as the liquid must/wine becomes richer with anthocyanin. At the same time the tannins are better extracted as the water/alcohol solution becomes more alcoholic and their extraction rate grows slowly with the fermentation. (Tannins have greater affinity to ethanol than to water.) Separation of the skins from the wine, namely pressing, finishes the extraction process. The timing is therefore very crucial in this matter. Short maceration, close to the end of regular fermentation (about 6 – 8 days) will leave the wine with almost its maximum color components, because these components are the fastest to be extracted. The tannins at this stage are not yet in equilibrium with the wine solution. On the other hand, an extended maceration which is practiced up to (20 – 30) days beyond finishing fermentation, will increase the tannin content and will decrease the color component in the wine.

Because the skins are pushed up during the fermentation by the CO_2 evolving from the fermenting must, a cap is formed on top of the must, with two main results: there is no contact between the juice and the skins, and heat cannot get out, which elevates the cap's temperature very much. The cap has to be treated to release it from being at the top, and to enable skin/juice contact, and to release the high temperatures. It is done by various methods such as punching-down (manually or mechanically), or by pumping-over the liquid must/wine on top of the cap to wet it and flow down through it. Each of the methods is done few times a day until skin separation.

In order to facilitate the extraction rate, especially in big scale wineries when time is very short at harvest, a special rotating tank was developed. In these tanks, instead of using the traditional punch-down method or the pumping-over method to wet the cap and maintain the contact of the skin with the wine, the whole tank rotates (horizontally) several times a day in both directions, with spiral blades inside the tank to break the cap, so that the skins are in good contact with the fermenting wine. The operation is planned to take a few days (3 – 4) in order to save time, and then the wine is pressed and finishes the fermentation in ordinary tanks. The idea behind these expensive tanks is that in a short time (up to 4 days) it is possible by a mechanical device to extract the phenolic at the same final content as would take (8 – 12) days in regular tanks. The only problem is that the final concentration depends not only on the diffusion rate between the skin/must contacts, but also on the dynamic equilibrium state between the two phases. Once it is established, no more phenolic can move out from the skins. So the rate of extraction may be higher with this technique, but a limit is set up by the equilibrium state which is time dependent. In any case, and in any technique that is used, the final maximum concentration will be defined only by reaching equilibrium. After vinification is over, the final wine made by this technique (after 3 – 4 days of maceration with good skin/juice contact) is still lower in phenolic content than one would expect, due to the shorter maceration time.

Other attempts which are used in the industry to enhance tannins and anthocyanin in wine are:

Cold soak – of the must before fermentation for a few days;

Long maceration – after fermentation for over a month;

Thermo-vinification – namely, pre-heating of the whole grape cluster before fermentation within a range of (40 – 80) °C to break the skin cell tissues;

Freezing – of the berries before fermentation, also to break the skin cell tissues;

Pectolytic-enzyme – treatment to break the skin cell tissues;

And lastly, *juice bleeding (saignée)* – namely, to separate part of the juice from the must to increase the skin/juice ratio.

Many studies have been done on these techniques; some of them using the same technique got contradictory results. But the overall conclusion is that right after fermentation, in some of these treatments, the concentrations of anthocyanin or tannins were higher to some extent, but afterwards in the long run in the finished wine, practically no advantage was observed, including wine quality measured by sensory evaluation. For all practical purposes the only method that enhances the phenolic content in the finished wines for the long run is higher temperatures of fermentation.

A decent and basic method to enhance the color intensity of wine may be to blend it with a small portion of high-color-intensity varieties, such as Syrah, Petite Sirah, Petit Verdot, Tannat, and others.

On the opposite side of heavy, high-tannic wines, are the light, fresh, and young-consumed wines. These wines are produced from the appropriate grape varieties, and with a short skin contact treatment, to minimize tannin content while still having a good red color. The lightest red wine production is done by a different technique called "carbonic maceration." This is a very special fermentation technique, which utilizes the ability of enzymes naturally present in grape berries to transform some small amount of sugar into ethanol. The process is eventually stopped by the accumulating alcohol inside the berries, which poisons the enzymes in the cells at about 2% ethanol. CO_2 is also formed in this intracellular "auto fermentation" (without any microbial activity). The enzymatic activity is enhanced at high levels of CO_2 (over 50%), or at low oxygen levels (below 5%). This technique was developed and is practiced mainly in France, in the production of light red wines with low tannin, primarily fresh-style wines intended to be consumed young (*Beaujolais Nouveau* wine). In this technique the grapes are uncrushed, and the whole berries (with stems) are placed in a tank filled with CO_2 gas to decrease oxygen contact during the process. This is the origin of the name "carbonic maceration." The berries are partially crushed by the clusters' weight, and juice flows to the bottom of the tank, where natural yeast fermentation takes place. After several days (up to a week or two) the grapes are pressed and left to finish their yeast fermentation to dryness. By such vinification one can get good-colored, softer red wine, with a characteristic fresh aroma, which is caused by a series of volatile aldehyde compounds formed during this technique.

To summarize this subject, these very general principles should be remembered: the pigments are extracted more quickly than the tannins, the must should be checked during fermentation, and differentiation should be made between the color intensity and the tannin level in accordance with the kind of wine desired.

C. Free-Run and Press-Run

1. General View

Pressing of grapes is done on the unfermented must for white and "blush" wines production, and during (or after) fermentation of red wines. The juice (or wine) that comes out of the press by using no pressure or just minimal pressure is called *free-run*, whereas the pressed juice (or wine) is called *press-run*. The unfermented free-run must is about (85 – 90) % of the total *extractable juice*, whereas the fermented free-run wine is

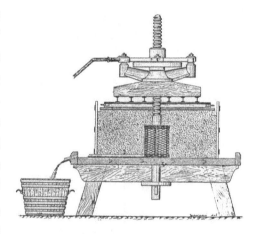

somewhat higher, about (90 – 95) % of the wine volume. The difference is due to the fact that during fermentation the pulp cells die, and the cells' membranes are not active anymore, making for an easier flow of the cells' content.

The constituent concentrations of the free and pressed run are different. In press-run (of must or wine), there are more phenolic compounds (pigments and tannins), less total acidity, higher potassium concentration, and therefore higher pH. In pressed wine, there is also higher volatile acidity and higher "vegetal" or "green" aroma. All these parameters are related to the pressure level exerted during pressing and the type of press used. Based on all parameters, better quality wine is made from free-run, whereas the use of press-run has to be considered according to its particular quality. It is advisable to separate the press-run from the rest, and process it in a different tank. If it has good quality, it can be blended later with the free-run wine in such a proportion as to get the highest quality possible. The press-run gives the winemaker a possible tool to enrich his wine in color (in red wine), tannins, and complexity.

Grape juice (like many other fruit juices) contains *pectin*. These polymeric carbohydrate compounds tend to create colloid coagulation in the wine, which may later cause an instability factor. Pressed juice contains more pectin than does free run juice, because pectin is found mainly in the skins. Hydrolyzation of the pectin during juice processing will prevent this potential problem. The hydrolyzation can be done by *pectolytic-enzymes*, which are commercially available. Wine or grape juice treated with pectolytic-enzymes will clarify faster and will be more easily fined. Its use is recommended for either white or red wines. The enzymes are tolerable to must conditions. Its pH optimum activity is between (3.5 – 5.0). Its optimum temperature activity is at 50°C, but it is active as well between (5 – 60) °C. Sulfur dioxide can act as an inhibitor to its activity, but only above SO_2 concentration of 500 ppm. The enzyme is also tolerable to all wines' alcohol levels, so it can also be used for finished wine. The practical range of enzyme addition is (2.5 – 5.0) gr/HL of must.

2. Presses

Presses can be divided into two major categories: batch presses and continuous presses. In the first category, the oldest and simplest press is the vertical basket press, traditionally made of wood. The pressure (up to 5 kg/cm^2) is exerted either manually by a screw jack or mechanically (hydraulic, pneumatic, or electrically), compressing the pomace downwards. When the pomace is pressed, the juice runs out through the basket rim  and is collected from its base. The basic problem with this kind of press is that the pressure is exerted on the pomace block, leaving some of the liquid trapped inside with no way out. The pressed "cake" is always wet after pressing. In other words, the pressing yield is low. Also, unloading the dense pomace after pressing, namely, breaking the compressed "cake," is difficult and requires hard work.

A modification of this kind of press is a horizontal press, made of stainless steel. The pressure is exerted by two plates moving from the edges of the cylinder, pushing the pomace toward the center and compressing it. The pressure is released from time to time, while the cylinder rotates slowly to break down the pomace "cake," enabling more homogeneous pressure, and allowing easier releasing of the trapped liquid.

A much more sophisticated and modern version of this press is based on inflating an air sac (bladder) mounted to the side of the cylinder. In the older type of inflating press, the air sac was centered in the drum, pushing the pomace towards the walls. In modern presses of this type, the bladder is mounted on the sidewall of the cylinder. The pressure in the sac is built up gradually in cycles, where each cycle is composed of the following: pressing the must gradually while rotating the drum, holding the pressure at its maximum value and releasing the pressure while reversing the direction of rotation (to break the pomace "cake"), then pressing gradually again while changing back the direction. The pressing time and the reverse rotation time in each cycle are set prior to operation. Normal cycling time is 2 – 5 minutes.

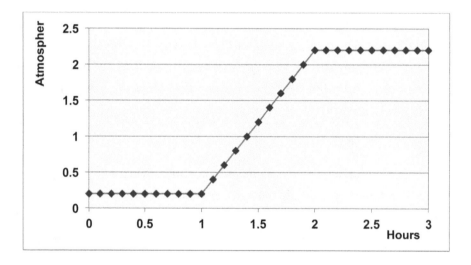

Typical pressure profile in the bladder press. Each point on the graph is a mini cycle.

The press operates in three cycling blocks according to the maximum air pressure in the bladder. The number of cycles in each block section is determined by the press operator and is automatically controlled and programmed. In block I, at low pressure (about 0.2 bar), the number of cycles is about (10 – 15). In block II, the pressure is incrementally increased at each cycle by 0.2 bar, up to the maximum pressure, which is pre-set by the operator (about 2 bars). In block III, the pressure in each cycle is set to the maximum, for about (5 – 10) cycles.

The total pressing time (including loading, pressing, and discharging) is about (2 – 4) hours. Because of the relatively low pressure exerted on the pomace and the gentle mechanical treatment of the skins, the solid content in the juice is relatively low, (2 – 4) %. This is most important in white wines, because it allows one to start the fermentation with relatively clear juice, which is crucial in producing quality white wines. The distinction between the free-run and pressed juice is determined at a certain pressure level during the pressure slope of the curve. Free-run juice is determined at a certain pressure (say 0.4 atm.) and from that point the rest of the must from the press is considered as press-run. Another option to decide when to separate the free-run from the pressed juice is by tasting it frequently as it comes out from the press. The main parameter for decision is the change in the acidity taste. The pH gradually gets higher with a flat taste.

In a case where the skins are thin and slippery, the screens inside the drum are usually plugged, causing difficulties in the press operation. (This phenomenon is typical, for example, in Semillon or Sauvignon Blanc.) A good practice in such a case is to add some portion of the stems (from the destemmer/crusher machine) to the must when loading the press (layer of must, layer of stems and so on). The existence of the stems inside the "cake" prevents plugging of the screens and allows the juice to flow easily.

All types of presses described above are batch types. They must be loaded and discharged at each batch. The batch quantity may run from about 2 HL (small basket press) up to (10 – 12) tons in the bigger horizontal air sac presses.

The second category of presses is the continuous press, which is based on an "infinite Archimedes screw" that presses the pomace slowly to an outlet whose size is controlled. This enables the operator to set the exerted pressure on the pomace as desired. The advantage of this type of press is that it is continuous, saving the loading and discharging operation. The main (and very serious) disadvantage of the screw-type press is that it causes tearing and breaking of the skin tissue, leading to more solids in the juice. A continuous press is not recommended for pressing white must, if recommended at all.

3. Lees

The lees are the solid particles suspended in the must after crushing and/or pressing. This term is also used to describe the sediment that settles before the first racking or after fining, and contains mainly yeast cells or fining particles. When leaving the white must to settle for (12 – 24) hours before fermentation (depending on the particle size and quantity), the settling lees create a dense layer (sludge) of variable thickness, at (5 – 10) % of the liquid volume. The juice above the lees layer can be racked off through the racking valve in the tank, but the only way to save the juice trapped in the lees layer is to extract it by mechanical force. This can be done either by lees filter, or by centrifuge. A lees filter can be very effective in clearing the juice from its solids. The filter contains (20 – 40) supporting plates, which support thick cloth pads. The plates (40x40 cm) with the pads are packed together tightly in a row, at very high mechanical pressure or hydraulic pressure of about (300 – 500) bars. At the entrance of the filter there is a strong pump, which can exert high-pressure forces on the lees to flow through the pads. To facilitate the filtration of the dense sludge, it is mixed well with diatomaceous earth (DE) before entering the filter. The DE is a finely ground mineral powder which increases the surface area of the filtering pads. This enables the lees to combine during filtering in a multi-layer texture with the DE, rather than in one layer on the cloth pad surface without DE. The mixing ratio between DE and the juice is about (2 – 4) Kg/HL, depending on the density of the lees. More DE is needed when the lees density is higher. The first batch of mixing can be (3 – 4) Kg/HL. After this batch is filtered, and a new batch of juice is mixed with DE, the DE/juice ratio can be cut to half of the first batch. If a third one is prepared, the ratio can be reduced to a quarter of the first addition.

At the beginning of filtering, the flow is easy and no pressure is developed. The first batch of the mixed juice and DE is used to build the DE "cake" between the pads, protecting them from being plugged by the fine lees particles. As the volume of liquid flowing through increases, more DE/lees particles are deposited, increasing the flow resistance. Pressure gradually builds up by the "cake," increasing resistance. The maximum workable pressure in most filters is about 10 bars. The pump is pressure controlled, and stops when reaching the maximum pressure for a couple of seconds, until the pressure drops down and it starts again. When the pumping intervals exceed (10 – 15) seconds, the filter is plugged and filtration should be stopped. The filtering capacity of a lees filter with (40x40) cm plates is about (30 – 50) liters/plate, so a typical lees filter with 20 plates can filter (600 – 1000) liters of must before it is plugged. Some useful hints for lees filtering operation:

At breakdown of the filter plates, the shape of the "cake" reflects the quality of the filtering. A "good cake" (dry, dense, and easy to remove) shows the proper ratio between DE/lees and good pad condition. When the cake is soft (not homogeneous and sticky), it means that the filtering was not efficiently done, the DE/lees ratio was too low, or the cloth pads were almost plugged with lees particles by early use of the filter.

Good washing of the pads with water after each use is necessary to prevent microorganism growth. Also, after a couple of filterings, the pads get partially clogged, and it is necessary to take them off the plates to machine-wash them.

Before closing and compressing the plates for new filtering, wet the cloth pads with water.

Check the edges of the pads before closing the plates, to assure that no bent pad is caught between the plates. If this happens, the filter will leak when the pressure rises.

The first (20 – 40) liters of filtered juice have an earthy taste (from the DE). Consider discarding it.

From time to time, open the high-pressure pump valve to release the DE that accumulates there. And finally, some remarks: In general, the quality of the lees filtered juice or wine is poor. When dealing with the lees left over after racking (yeast lees, fining lees), generally the volume of the lees layer is small, and it is sometimes better to simply drain it off. Also, the DE dust is not healthy to breathe, and it is highly recommended that the operator wear a dust mask while mixing the DE with the must.

D. Must Corrections
1. Acidity

When we previously discussed the acidity of must and wine, it was noted that the two parameters are connected, namely, the acidity and pH are related but not in a simple linear relation. The distinction between *total acidity* and *titratable acidity* was made in the previous chapter. Also the terms pH, pK_a and buffer capability of wine were defined. (The reader is referred to section A in chapter I.)

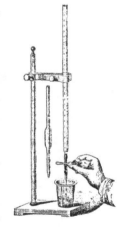

Some words concerning the buffer capacity of must and wines: the buffer capacity, which is the response of the pH changes to addition of strong base (or acid), depends on the titratable acidity of the weak organic acids in the case of wine, *and* on the main cation concentration in wine, namely, of potassium in this case. The higher the titratable acidity and potassium concentration, the higher is the buffer capacity (but not linear relation). A titration curve of a typical must, and of a strong inorganic acid for comparison, is shown in the following figure:

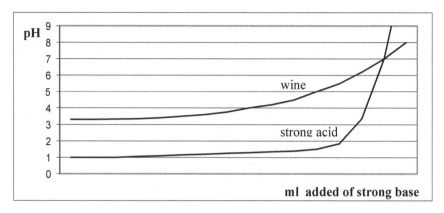

Titration curve of a typical wine (or must) and a strong acid by a strong base.

Note the sharp change in the strong acid titration curve at the neutralization zone, compared with the wine curve, which is quite smooth and continuous. The titration end point is of course at pH = 7, but it is difficult to define it exactly when done in wine. When a chemical indicator is used for determining the end point by changing its color (instead of by reading the pH-meter), the end point should then be at the pK_a value of the indicator, which is not necessary at pH = 7. The most-used indicator in white wine titration is *phenolphthalein* (changing its color from colorless to red at the end point when titrating the acidic wine by base solution), with the end point of its pK_a value at pH = 8.2. For convenience and for common practice, this pH was chosen as the end point of TA measurements in the wine industry, even when the titration is measured by a pH-meter. This is the case when measuring the TA of red wine, where it is impossible to use an indicator because the wine's red color masks any other color. So the end point is therefore determined by a pH-meter, at pH = 8.2.

The buffer capacity of must is in the range of $(4 - 6) \times 10^{-2}$ equivalent/L per one pH unit, meaning that in order to change the pH of 1L of must by one pH unit, it is necessary to add $(4 - 6) \times 10^{-2}$ equivalents of strong base or acid.

Measuring and knowing the TA and the pH values of the must is the basis for considering whether adjustment of these values is needed, for two reasons:

(1) To adjust the sour taste of the wine. It may be either too sour (high acidity of the must), or flat (low acidity). The adjustment is based according to the desired type of wine.

(2) Adjusting the must pH will affect its stability to chemical and microbial changes. This case is mostly relevant in wines with too high a pH.

The *sour taste* is more dependent on the concentration of acids (*total acidity*) than on the proton ion concentration (expressed by its pH). As a very general guide, the preferred TA and pH values of the white must would lie in the range of TA = (6 – 9) gr/L, and pH = (3.1 – 3.4) respectively, and TA = (5 – 7) g/L and pH = (3.3 – 3.8) for red must. In sweet or semi-sweet wines, the acidity should be at the higher end, and the pH at the lower one, in order to balance the sweet taste. It should also be kept in mind that the TA and the pH will change during the wine processing. The TA will become higher after fermentation by the addition of new acids produced as by-products of the fermentation (e.g., succinic acid, acetic acid, malic acid, and others). Also the acidity might be reduced by malo-lactic fermentation and by cold stabilization (to be discussed later). The pH too will change its value in the finished wine. The acidity issue can be divided therefore into two categories: *deficiency* and *surplus*, or in other terms, *too high* pH and *too low* pH.

Too High pH

The more severe and common case in white must is the lack of acid rather than the surplus of it, especially in warm regions. The advantages of reducing the pH and increasing the acidity are: (a) more controlled and steady fermentation, (b) reduction of the yeast autolysis at the end of fermentation, and (c) better inhibition of bacterial spoilage. It also preserves the flavor and the aroma of white wine, and it is necessary for the balance and general quality of white wine. Too high pH is also not desired in red wines, mainly because of microbial stability problems, reduction in color intensity, and flat taste.

The acid correction should be done before starting the fermentation, for the above reasons regarding fermentation, and because any major correction later on will greatly affect the wine's stability. One further very important point: when the major correction is made later on the wine, there is greater chance that the correction will affect the perceived sour taste, giving the impression that something is not right.

The techniques that can be used for acid correction when in deficiency are: blending with high acid must, addition of acids, and ion exchanging.

(1) *Blending* with other must (high acidity and low pH) is the best natural and elegant method, whenever possible. However, in many cases, it is not practical. It may be quite difficult to find the right grapes with the desired pH and total acidity,

at exactly the time of crushing and starting fermentation. There are also other factors that should be considered in blending, which are not necessarily dependent only on the pH and TA, especially when blending with another variety. Such blending may also change the quality and character of the wine.

(2) *Acid addition* is the most practical method to correct deficient acidity. In this case the only expected change is solely the wine acidity and pH. The addition options are to use tartaric, malic, and citric acid; each has its advantages and disadvantages.

a. *Tartaric acid* is the strongest of the three (lowest pK_a), and less of it would be needed than of the others to get the same result. On the other hand, a large part of the added acid will precipitate later during cold stabilization. However, by cold precipitation of potassium-bitartrate salt, the potassium concentration is reduced, which leads to a lower pH value. If high pH is the major problem of the must, the addition of tartaric acid and precipitating its surplus potassium by cold stabilization will improve the pH value of the wine considerably. (For more details see the Cellar Operation chapter, section B.)

b. *Malic acid* is the principal cause of acid deficiency in must during the ripening period, due to respiration. The addition of malic acid into the must can be looked at as a way of restoring the original balance between tartaric and malic acids in the grapes. Malic acid, which is less acidic than tartaric acid, will not precipitate in cold temperatures, but may be transformed into lactic acid by malo-lactic fermentation if it happens. When the commercial DL-malic acid is used to acidify the must, only the L-malic acid isomer is consumed during the malo-lactic fermentation, leaving the D-malic isomer (50% of the added DL acid) untouched. So, even if the malo-lactic fermentation takes place, the addition of malic acid will partially contribute to the total acidity. In wines in which malo-lactic fermentation is not desired, and precautions are taken to prevent it (mainly in white wines), the advantage of its addition is clear.

c. *Citric acid* can be added to expand the acid taste and to prevent possible iron haze (by complexing the iron). However, during ML-fermentation, it may be partially converted into acetic acid. So, citric acid can be added safely only when ML-fermentation is inhibited because it is not desired in that particular wine, and measures have been taken to prevent it. Otherwise, it can be added after the ML-fermentation is completed. There is a legal limit of its content in wine, e.g., in the U.S. it is 0.7 gr/L, which restricts the use of this acid only to small quantities.

The amount of acid needed to correct the acidity deficiency depends on the TA, the pH, and the buffer capacity of the must. In most cases, as a rule of thumb, when using tartaric acid at a pH range of (3.6 ± 0.3), every 1 gram/liter added will lower the pH by about (0.1 ± 0.02) pH units. The minus sign of the deviation $(- 0.02)$ is in the musts where the initial pH is smaller than 3.5 value (pH < 3.5), and the plus sign $(+ 0.02)$ is when the initial pH is higher than 3,5 value (pH > 3.5.) For malic or citric acid, each 1 gram/liter added will lower the pH by about (0.08 ± 0.01) pH units, at the same pH range as above. This is a general guideline. Before adding the acid to the must (or wine), one should try to determine the right quantity per liter in the laboratory, in order to get the optimal TA and pH according to the case.

In the rare cases where the pH is too high, and the TA is also high (caused mainly by high potassium content), addition of more acid in order to lower the pH will result in a very harsh taste of artificial added acidity. This is a very difficult case. In such a case there is the option to use $CaSO_4$ (gypsum) to reduce the pH in combination with tartaric acid. This is in accordance with the European Market regulations, which allow maximal addition of 1.5g/L of tartaric-acid to must and 2.5g/L to wine. The regulations allow the use of $CaSO_4$ in wines with a limit that the residual sulfate content, expressed as K_2SO_4, will be less than 2.5 g/L. In the U.S. the maximum legal content of sulfate is 2g/L. Gypsum is practically not dissolved in water and it is used as fine powder, mixed well in the must by pumping. This method might solve the difficult cases of must with high TA and high pH. In the case of red must with TA = 7g/L for example, and if tartaric-acid is added (say to reduce from pH = 3.95) in order to reduce it to a reasonable value (say, pH = 3.7 – 3.8), the TA after addition might be too high in *sourness* aspect. By using $CaSO_4$ it is possible to reduce the pH while leaving the TA nearly at the same level.

This method is not as common, because of the slow precipitation of calcium ions as calcium-tartrate from the wine, and a bitter aftertaste which sometimes follows this treatment. But facing the options of having must with a really serious problem of *high pH* and *high TA*, it is worth considering. And if this method is chosen, it should be done to the must before fermentation to minimize the side effects, and then only after a lab test on a small volume to find the optimal $CaSO_4$ and tartaric acid additions. (For more details in this matter the reader is referred to *Concepts in Wine Chemistry* book, chapter VII, section C.)

(3) *Ion exchange* method enables one to exchange the potassium cation with hydrogen, and is a good technical solution to the above case of high pH and high TA. The ensuing exchange is very efficient, and almost all the potassium is replaced by hydrogen ions, lowering the pH substantially. Only a portion of the wine is transferred through the ion exchanger column being stripped of its potassium ions and then blended with the rest of the wine. The result of such blending will lower the pH of the total wine batch.

Too Low pH

Must de-acidification is the other side of the problem, where the must TA is too high (above 9 gr/L) and the pH is too low (below 2.9 – 3.2). This problem is usually the result of unripe grapes that had to be harvested because of some viticulture difficulties, such as weather conditions, vine diseases, or bunch rot. The problem is quite common in Germany, parts of France, Switzerland, and other cool regions. The taste of the wine made from such must is very sour, tart, and unbalanced. Low pH will also inhibit ML-fermentation when it is most desired to reduce the acidity. In order to correct the excess acidity and to increase the pH level, some measures can be taken, namely, blending with low pH must, chemical treatment and malo-lactic-fermentation as follows:

(1) The best method is to blend with low TA and high pH must. Again, the same considerations mentioned earlier in this chapter on low acidity correction by blending are applicable here as well.

(2) Another way to reduce the acidity of the must (or wine) is by chemical treatment. In principle, the neutralization of the excess acid by a base is not good, because the salts formed will give the wine a salty taste and sometimes an off-flavor. It is preferable to remove the excess acid by precipitation with *calcium carbonate*, *potassium carbonate*, or *potassium tartrate*. These salts will react with the excess tartaric acid to produce hard soluble tartrate salts, which will precipitate and lower the TA. The reactions are:

Calcium carbonate:	$CaCO_3 + 2H^+ + 2HT^-$	===>	$Ca(HT)_2 + CO_2 + H_2O$
Potassium carbonate:	$K_2CO_3 + 2H^+ + 2HT^-$	===>	$2KHT + CO_2 + H_2O$
Potassium bitartrate:	$K_2T + H^+ + HT^-$	===>	$2KHT$

All products on the right side of the above equations precipitate with tartaric acid, reducing its concentration from the must. The quantities of calcium-carbonate or potassium-tartrate which have to be added, must be determined in the laboratory before addition. As a general guideline, experience shows that the following quantities upon addition to the must will increase the pH value by approximately 0.1 pH unit:

$CaCO_3$ (0.3 – 0.4) gr/L

K_2CO_3 (0.4 – 0.5) gr/L

K_2T 1.5 gr/L

(3) ML-fermentation is a good and "natural" treatment for reducing acidity in wine. In this fermentation, the malic di-acid is transformed into the lactic mono-acid, consequently reducing the concentration of wine acidity by half of its malic acid concentration:

$$HOOC – CH_2 – CHOH – COOH =====> CH_3 – CHOH – COOH + CO_2$$

Also the pK_a of lactic acid is higher than the pK_a of malic acid, which means that lactic acid is weaker than malic (besides reducing the malic acidity by half). A major problem in this method is that when the pH is below (3.1 – 3.2) it is difficult to start the ML-fermentation. In such a case where the pH is lower than this value, it can be brought up to above (3.2 – 3.3) by other methods (such as carbonate addition, or blending) followed by inoculation with ML culture.

2. Sugar

Many reasons may cause a deficiency of sugar concentration in the grapes at harvest time. They include weather conditions, nutrient and water deficiency, and all kinds of diseases which end up in unripe grapes. When this happens, leaving the grapes on the vine for further maturation and delaying the harvest may not improve the sugar accumulation. In certain cases it may actually cause damage to the grapes. With unripe grapes, there is not enough sugar to ferment into the normal alcohol concentration range of wine. Usually such grapes also contain excess acidity. There is nothing wrong, from the quality aspect, with making sugar or acid corrections. However, because these grapes are unripe, the quality may suffer from a more severe problem which cannot be corrected, specifically, not enough varietal flavor and aroma. In cold regions (Germany, north of France, Switzerland), the addition of sugar is necessary in certain seasons, and is legal. The regulations usually require a certain minimum of natural sugar content around (15 – 16) B° and notification to the authorities on the sugar addition needed (*chaptalization*). In other regions, for example California, it is absolutely forbidden.

The sugar addition can be done with cane sugar (sucrose) or with grape concentrate (about 70% grape sugar). When sucrose is added, it is naturally hydrolyzed to mono-sugars by inversion enzymes produced by the yeast.

The quantity added depends on the sugar level in the must and on the desired alcohol concentration. The following formulae are useful in calculating the amount of sugar needed in each unit of sugar terms:

0.92 Kg sugar/HL ====> 1 Brix

1.65 Kg sugar/HL ====> 1 Baumé

2.10 Kg sugar/HL ====> 10 Oechsle

For example, 0.92 Kg of sugar added to 1 HL (100 L) of must will increase the Brix by 1 unit. If cane sugar is used, it is best to dissolve it in boiling water to make very highly concentrated syrup before adding to the must. The volume needed can be calculated from the syrup concentration.

To prepare concentrate syrup, boil 25 liters of water, and add 50 Kg of sugar while mixing. After a few minutes the syrup will clear up. The syrup volume will be about 55 liters at roughly 66 Brix (the volume of 1 kg sugar solution is 1.1 L). To measure its exact concentration, dilute a sample by 5 (1 volume of syrup and 4 volumes of water) and measure its Brix with hydrometer (16 – 24) B° range. The syrup concentration will be the Brix reading multiplied by the dilution factor (x5). The Brix unit is grams of sugar dissolved in 100 grams of solution. The volume that has to be added can be calculated by the following example. Assume that the syrup is 66°B and you want to add 3 more Brix to the must. Then add: 3x0.92x1.1 = 3.0 liter of syrup/HL of must. Or, if you want, you can weigh 3x0.92/0.66 = 4.2 kg of the syrup/HL of must.

It is recommended to add the syrup while it is still warm at (40 – 50) °C, before the sugar recrystallizes. It should also be noted that dilution of some few percent of the must with water is caused by this operation. If grape concentrate is used, the producer of the concentrate indicates the sugar concentration in B°, and the same calculation can be made.

The addition of sugar has to be done at the beginning of fermentation, when the yeast is vital and highly active. If added at the end of fermentation, it can sometimes get stuck, and it will be difficult to achieve dryness.

E. Cooling and Temperature Control

The temperature is one of the external circumstances which have the greatest share in influencing the act of fermentation. It has been considered, that a heat of about the 54° of Fahrenheit's scale is that which is most favorable to this process. There is nevertheless some latitude to be allowed; but in a temperature either very cold or very hot it does "not take place at all."
–From *Remarks on The Art of Making Wine* by J. Macculloch, 1829

1. Fermentation Temperature

Fermentation is a heat source process. Yeasts ferment sugar in order to get their energy out of it. This anaerobic process has a poor yield of energy, and the yeast uses about 57% of the available energy of transforming sugar into ethanol. The rest is released mainly as heat (and some as mechanical work of the CO_2 bubbles). Theoretical calculations show that by fermentation, each $1B°$ releases about 1.15 Kcal as heat. For example, one liter of must containing $23B°$ will release 26.5 Kcal. In other words, if the fermenting tank is thermally isolated without releasing its heat, the temperature of the fermenting wine could rise (theoretically) by 26.5°C from its initial temperature. Practically, this is not the case. First, if the temperature rises to above 36°C, the fermentation will stop. Second, natural heat losses through the container's walls reduce the expected theoretical temperature. This reduction depends on the total volume of the must (or more accurately, on the *walls' surface/ volume ratio*), on the container material (cement, wood, or stainless steel), and on the environmental temperature (temperature difference between the fermenting must and the surroundings). The natural heat loss is not enough in most conditions, and cooling is necessary. Therefore, controlling the temperature during fermentation is one of the most important factors in good winemaking. The range of temperatures in which yeasts are active and can ferment sugar is between $(10-35)$ °C or $(50-95)$ °F. At the higher temperature range, the fermentation starts faster, but as the alcohol concentration increases, it slows down. At around 35°C it might stop, leaving some residual sugar unfermented. At low to moderate temperatures, the fermentation starts slowly, proceeds more moderately, and generally will go on to dryness. The time lapse between inoculation and the first signs of fermentation (lag period) can be from a few hours to a few days, depending on the temperature and other factors (which

49

will be discussed in the next chapter). The alcohol formed during fermentation is an inhibitor for yeast growth, and its inhibition effect is greater at high temperatures. It also evaporates faster through the CO_2 bubbles at higher temperatures, which makes the alcohol yield higher at low temperature fermentation. The variation in alcohol content can be as much as 1% absolute alcohol difference if the fermentation is carried on at temperatures of 30°C and 10°C. Also, at lower temperatures, the fruitiness of the grapes is better preserved, by reducing evaporation of volatile aroma components from the must. Also, the volatile acidity level at low temperature fermentation has been found to be lower than at high temperatures.

For all the above reasons, and for practical considerations, it has been accepted that the preferred fermentation temperature of white wines is between (8 – 14) °C or (46 – 57) °F. This is also true for rosé wines, and white wines made from red grapes ("blush wine"), as they are considered white wines. In red wines, on the other hand, the fermentation temperature should be higher, between (22 – 32) °C or (72 – 90) °F, for two reasons: color and tannin extraction are better, and the fruitiness compounds of red wines are less volatile at these temperatures than those in white wines.

2. Temperature Control

In order to carry on the fermentation at the desired temperature, one should control the temperature by means of a mechanical cooling system. This modern technique uses cooling liquid (usually ethylene glycol) flowing in a cooling jacket around the tank. The liquid is cooled by a cooling system, which must have the cooling capacity needed for all winery functions. The tank's jacket is generally built as one or two strips at 2/3 and 1/3 of the stainless steel tank's height. The fermentation heat flows from the tank's walls to the cooling liquid, and then away to the cooling machine. The temperature can be controlled individually at each stainless steel tank by a temperature sensor placed in the tank. The sensor should be long enough to read the liquid temperature far away from the walls, where the temperature is the coldest.

Because there are two temperature gradients in the tank, one radial, from the center toward the walls, and the second from top to the bottom of the tank, the temperature control does not necessarily read the true temperature of the fermenting tank. The middle temperature to control is probably in the half height vertically, and (15 – 20) cm horizontally from the wall of the tank.

During fermentation the two perpendicular gradients are reduced by the fermentation turbulence and by the pumping-over of the must in red wine fermentation. In old European wineries, the fermentation cooling was done simply by dripping cold water on the outside of the tank's walls, with the flow rate of the water serving as the temperature control. Another method is to use an air-conditioned room, which is much less efficient, because of the low heat capacity of air. In barrel fermentation (of white wine), this is the only method to control the fermentation temperature. In red wines, the fermentation may take between one to two weeks, whereas in whites, it may take three to six weeks to finish. In white wines, when the fermentation seems to be slowing down or even stuck towards the end of the process, it is advisable to consider raising the temperature slightly to about (15 – 18) °C in order to accelerate the completion of the fermentation. After fermentation is over, and during the rest of the winemaking process (racking, fining, barrel aging, etc.) up to bottling, the temperature should be controlled at (12 – 15) °C for whites and (20 ± 5) °C for reds (preferably at the lower range).

When quick cooling is needed (in white must), when the must has to stay on the skins for several hours at a cool temperature, or just to start fermentation at low temperature, there are cooling machines that allow reduction of the must temperature by (5 – 15) °C at a very high flow rate through a heat exchanger. In any fast cooling with a heat exchanger, it is highly recommended to pass the must through the cooling system to another tank, rather than to pump it back to the same tank. By returning the cooled liquid back into the same tank, the temperature gradient between the cooling system and the liquid gradually gets smaller and smaller, with lower and lower cooling yield. When the liquid is transferred to another tank, the gradient remains constant, making the cooling much quicker and more efficient.

CHAPTER III

FERMENTATION

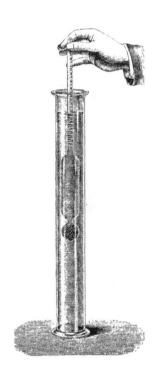

CHAPTER III: FERMENTATION

A. General

Fermentation is the heart of winemaking. The goal in this process is to convert sugar into ethanol in a way that produces a minimum of undesired byproducts, while preserving the maximum natural aroma and flavor of the fruit. Ideally, the fermentation will produce ethanol to the desired concentration range and will develop the unique character of the grape variety, its appellation and wine style. The basic expression of the fermentation process is given by the overall formulation:

$$(4) \qquad \underset{MW = 180}{C_6 H_{12} O_6} =====> \underset{MW = 46}{2CH_3 - CH_2 - OH} + 2CO_2$$

MW is the molecular weight of sugar and alcohol, and we can learn from the equation that one mole of sugar (180 gram) will theoretically produce two moles (2x46 = 92) of alcohol, which is **51%** of the sugar weight. Actually about 5% of the sugar is consumed to produce many by-products such as *glycerol*, *succinic-acid*, *lactic-acid*, *acetic-acid*, *2,3-butanediol*, and other products. Also, about 2.5% is consumed by the yeast as carbon source, and about 0.5% is left over as unfermented residual sugars. A total of about 8% of the B° is not converted into ethanol. This makes the total weight yield to be 51 x 0.92 = **47%** which comes to 180 x 0.47 = **84.6 gram of ethanol**. The volume of that amount of alcohol is the weight divided by its density (ethanol density at 20°C is d = 0.789 g/ml), therefore 84.6 / 0.789 = **107.2 ml**. When this volume of alcohol is diluted with water to make one liter of alcohol/water mixture, the total volume contracts by 0.7%. (This means, for example, that if you take 100 ml of pure alcohol, you should have to add 907 ml of water, which after the mixing will finally become one liter of 10% alcohol v/v). So the alcohol concentration in wine made from 180 g/L sugar would contain 107.2 x 1.007 = **108 ml** alcohol in one liter of wine which is **10.8% (v/v)**. In Brix units (where about 95% of the total solids content in must is sugar), in order to produce 1% (v/v) of alcohol, the following Brix will be needed:

(180 /0.95 /108 = 1.75), namely, each **1.75 B° ===> 1% (v/v) alcohol**.

This formula, which is very useful in predicting the alcohol content, can also be expressed as the ratio of ethanol % (v/v) to the must's Brix by 1.0/1.75 = **0.57**.

The meaning of this number is that the potential alcohol content in a given must can be predicted by the quite good approximation:

% potential alcohol (v/v) = 0.57 Brix

The experimental value of this number was checked in various studies and was found to be in the range of (0.55 – 0.60), namely, it can be written as:

(5) **% potential alcohol (v/v) = (0.57 +/- 0.03) Brix**

This formula has the following limitations:

(1) The non-sugar solids which are part of the must's Brix depend on the grape variety, growing region, and state of ripeness. Grapes which are less ripe have higher non-sugar solids content, and hence, lower alcohol/Brix ratio.

(2) Ethanol yield depends on the fermentation temperature: the higher the temperature, the lower the yield, partly because of other by-products and partly because of ethanol evaporation at higher fermentation temperatures.

B. Wine Yeast

People have been making wine for thousands of years without any idea why or how the grapes' sweetness was transformed into alcohol. The process was not understood until Pasteur discovered in the mid-nineteenth century, that it is conducted by yeast. If the must is left to itself, the "natural" yeast found on the grapes and in the winery will start fermentation. Over many years, through research and trial and error, the more favorable yeasts for wine production were selected. Currently, about 150 yeast *species* are known to relate to winemaking, categorized into twenty different *genera*. The most important yeasts for wine production belong to the *Saccharomyces-cerevisiae* and *Saccharomyces-bayanus* species. These two species (the first name designates the *genera* and the second one is the *species* name) are sub-classified into different *strains* of yeast (all belonging to the same species), which are based on their unique characteristics for the winemaking process, such as:

a. Tolerance to various medium conditions, such as temperature, alcohol, and sulfur dioxide concentrations. Certain yeast strains are sensitive to cold medium, becoming inactive at low temperatures. This characteristic can be very useful for easily stopping the fermentation process before dryness to leave some residual sugar, by simply reducing the temperature. Other strains are tolerant to high alcohol content, and therefore can be used to ferment high Brix must to dryness, or to restart fermentation when it gets stuck toward the end. Generally, yeasts are not very sensitive to SO_2 at the concentrations used in winemaking during fermentation. However, there are some strains that are more sensitive to it.

b. Different by-products which are formed during fermentation. Some by-products are undesired, such as hydrogen-sulfide or volatile-acids. Sulfur dioxide is another by-product in this category.

c. Ability to ferment to the end, leaving low residual sugar.

d. Flocculation capability at the end of fermentation, which affects the yeast sedimentation after fermentation is over and hence the wine's clarification. This characteristic is important in sparkling wine production (besides other characteristics which are needed when dealing with the unique demands of second fermentation).

e. Kinetic of fermentation, namely, the rate of how fast the fermentation is proceeding. This is an important factor in winemaking and it is very different in various strains.

f. Flavor matching with various grape varieties. This characteristic is the ability of the yeast strain to develop during fermentation the release of the flavor components in the grapes from their sugar conjunction, so that they become more volatile and the varietal uniqueness can be expressed.

The commercial yeast strains of *Saccharomyces-cerevisiae* (also called *Saccharomyces-ellipsideus*) and *Saccharomyces-bayanus* are sold in dry vacuum packages. The loss of activity during storage of the dry yeast is temperature de-pendent. The activity loss after one year of storage at 20°C and 4°C is about 20% and 5% respectively. The viability of a fresh batch is about 20×10^9 cells/gram. The recommended addition dose of yeast to the must is about (200 – 250) gram of dry yeast/1000L of must to start the fermentation. Namely, the starting concentration is about $20 \times 10^9 \times 200/10^6 = 4 \times 10^6$ yeast cells/ml. At this concentration, the added yeast species cells are dominant in numbers over any "natural" yeast species, and very soon by their multiplication they will take over the lead of the fermentation. Modern winemakers use various species of yeast to best fit the wine type and style at hand. The use of a specific dominant yeast strain is most beneficial, making the fermentation more consistent, faster, and tailored to the winemaker's exact needs.

In contrast to this practice, there is the view that prefers to rely on "wild yeast" found in the vineyard and in the winery equipment. This choice lets the fermenta-tion start and proceed to finish "naturally." It is believed that by fermenting the wine with the various "natural yeast" strains, the wine will be more complex, with higher quality than by only one selected strain. When using "natural yeast" fer-mentation, control over the various parameters mentioned above is not at hand. The start of fermentation might be delayed, might not finish the fermentation to dryness, might develop various off-flavors, and in general the uncertainty which always is part of this process is higher with "natural yeast." For example, one of the most abundant strains found in the so-called "natural yeast" belongs to another genera, *Kloeckera-apiculata*, which is more active at low alcohol concentrations at about (4 – 5) % alcohol. It tends to produce higher concentrations of volatile acids during fermentation, and because it is dominant at the low alcohol stage, it has the advantage over other yeasts. By using cultured yeast we introduce about 100 times cells/cc than "natural yeast," so very soon they will be practically the only players in the game. And one should not forget that all cultured yeast strains are members in the whole community of the "natural yeast" that were isolated for our purposes.

We find it impractical to detail here the various strains available on the market. Instead, we recommend studying the professional material on the various yeast strains which can be obtained from the major yeast producing companies such as Lallemand, Red Star, Enoferm and others. In these materials one can find all the information needed regarding any yeast strain, its origin, and microbiological activity, sensitivity to various conditions, oenological properties, flavor characteristics, and recommended usage.

Killer Yeast

Killer yeasts are certain mutations of yeast which release into the medium in which they are living a lethal toxin to other yeasts of different strains. The killer yeasts are immune to their toxin, while other yeasts are either sensitive or unaffected. This phenomenon was identified in many wild yeast genera including the *Saccharomyces* one. The two yeast species, *Saccharomyces-Cerevisiae* and *Saccharomyces-Bayanus*, used in the wine industry, contain also the killer yeast mutants. In this respect all yeast strains can be classified in three categories:

Killers (K) which have the ability to kill other yeast strains of the same species, **Sensitive (S)** which are sensitive to the toxin produced by the killers, and finally the **Neutrals (N)**, which are not affected by the toxin and are not killers either. This specific character of any commercial yeast strain is listed among other ones by the yeast's producers. The Killers are named in the literature by a **K** letter followed by subscript number (K_1, K_2, etc.). The lethal toxin is a protein precursor, which when being absorbed by the sensitive yeast, interferes with the proton transport in the membrane cell, making it more permeable to protons. The results are lethal for the host cell. This matter is of great concern, because in certain cases the results are stuck fermentation. The literature reports on a wide range of killer yeast populations causing fermentation to halt. Study of these stuck fermentations showed that 90% of the yeast population was dead and the remaining 10% viable cells were killer yeasts. This phenomenon was first observed in 1963, and it has become a very disturbing issue in the wine industry since then, in various genera and species and in different locations around the world.

C. Fermentation
1. Alcoholic Fermentation

The major visible changes occurring during fermentation are a reduction in sugar concentration and an increase in yeast population. These two changes may serve to sketch the fermentation profile as shown in the following figure:

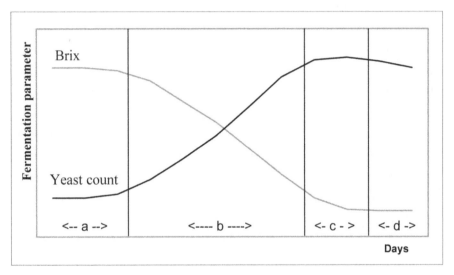

Schematic profile of fermentation by two parameters:
Brix reduction and yeast cell count

The yeast community multiplies itself by budding at a rate that is controlled by various media conditions such as: sugar concentration, temperature, nutrients, alcohol concentration and other by-products, and oxygen. The fermentation profile is characterized by four phases shown in the figure:

(**a**) *Lag phase*, where the yeast acclimates to the must conditions (high sugar level, low pH, must's temperature, and SO_2 if present).

(**b**) *Exponential growth phase*, where yeast multiply exponentially, reaching maximum population density (about 100 million cells per cubic centiliter). Sugar concentration in this phase declines exponentially as well.

(**c**) *Stationary phase,* where yeast density is at its maximum value and growth is practically halted. The fermentation rate slows down.

(**d**) *Decline phase,* where nutrients are scarce, the toxic by-products content is high, the viable yeast cells number is gradually decreasing, and the yeast mass is settling down.

In order to start a good and healthy fermentation, one has to add enough living yeast cells to the must. By using dry yeast, which is very convenient, the quantity needed is about $(20 - 25)$ gram/HL, which is about $(10^6 - 10^7)$ cells/ml. At maximum activity the yeast cell count reaches up to about 10^8 cells/ml. The population changes during fermentation can be followed by using a microscope with a magnifying power of about $(500 - 1000)$, allowing good vision on the yeast cells whose size is about 1-2 micron $(1\mu = 10^{-3}$ mm$)$ in diameter.

The yeast should be hydrated first by a water/must-mixture (10% must) at temperatures of about $(38 - 40)$ °C or $(100 - 105)$ °F, for about 15 minutes. On hydration, the ratio of the dry yeast to water is about 1:10 (1 Kg dry yeast to 10 liters of water/must mixture). Be aware that in $(10 - 20)$ minutes, the newly hydrated yeast may overflow from its container as a result of very fast fermenting activity. After hydration, it should be introduced directly into the must. Mixing is not necessary.

Regarding the temperature at the starting stage, there are no problems with red wines, as the temperature is about $(22 - 30)$ °C. For white wines, where the fermentation temperature is lower $(10 - 14)$ °C, it may occasionally take quite a long time before the fermentation starts. It is recommended therefore, to bring the must to a temperature of about $(15 - 17)$ °C, then to add the yeast and wait for good signs of fermentation (reduction of one-two Brix unit). At this point, the yeast population has grown enough to carry on the cold fermentation. Then reduce the temperature to the desired range $(10 - 12)$ °C. If the yeast in use is highly tolerable to cold fermentation (e.g., Prise-de-Mousse) the inoculation can be done directly at the cold fermentation temperature.

There are some procedural differences in fermenting white and red musts, because of the skin (cap) management needed in the reds, and the lower temperature needed for the whites. The volume of white must during fermentation should not be more than $(95 - 97)$ % of the tank volume, to leave some space for foam that might evolve. In red wine fermentation, the must volume should not be greater than (85 – 90) % of the tank volume, to leave space for the floating cap which is pushed up by the carbon dioxide gas.

The red must should be placed in a fermentation tank equipped with a pumping-over device, which can spray the pumped must from the bottom of the tank over the floating cap of skins. This pumping-over is necessary for color and tannin extraction, and should be done three to four times a day during fermentation, and then twice a day until pressing or until the cap settles down (in the case when pressing is delayed for long skin contact). Cap management can also be done mechanically, by punching down the cap in an open tank.

Other options are now available to treat the cap besides pumping over. One is a closed stainless steel tank that has a device to punch down the cap mechanically. The other option is a "rotor-fermenter," which is a big horizontal tank that can rotate very slowly around its axis. By doing so, the cap is well mixed with the whole must in regular time intervals. These tanks are quite expensive but they are efficient in extracting the skins, and consequently may save some time during the crushing period.

The fermentation rate has to be followed carefully from beginning to end by measuring the Brix reading and the temperature daily, and by smelling the tanks. If the fermentation rate does not follow the normal pattern (as shown in the above figure), or if something goes wrong, mainly the fermentation getting stuck, quick action should be taken to prevent serious problems which can develop later on and affect the wine quality.

Also, one of the most common problems during fermentation is what is called "stinking fermentation." Experienced winemakers can detect it even at the very early stages, by smelling the fermenting tank. The bad smell is caused by formation of hydrogen-sulfide (H_2S) produced by the yeast in certain circumstances. This phenomenon is quite common and is mainly caused by nitrogen deficiency in the must. When discovered at an early stage of fermentation, the best treatment is to add (100 – 200) ppm of diammonium-phosphate (DAP) to the fermenting tank. The bad smell may be dissipated in short time if discovered and treated early. If it is discovered at the end of the fermentation, it is of no use to add DAP, and the best treatment is to aerate the wine by racking. Aerial racking can be done simply, by pumping the wine from that tank to an open top of another one. If the H_2S has not yet been changed into mercaptan (by interaction with the alcohol), the aeration will be very helpful. When not treated in time, the hydrogen-sulfide can become a serious problem that can be treated only by fining with copper. (For more details see chapter VII, section D.)

Some wineries having old winemaking traditions put a piece of copper bar in the must container at the output of the press (in white wine). This brief contact with the copper bar dissolves some copper ions into the must which then precipitate as copper sulfide when the H_2S is formed, eliminating the stinky odor. This method is quite efficient, but has the risk of introducing too much copper into the wine, which may cause copper haze instability later on. Also, copper ions may provoke the yeast to produce more hydrogen-sulfide. So this traditional practice is not recommended. The easiest way to prevent or minimize this problem is first, to avoid any sulfur dust on the grapes by good vineyard management, and second, by addition of DAP to the must *before* fermentation at time of inoculation. The recommended quantity was discussed above. A final word on safety during fermentation:

Never add additional must to a fermenting tank from the bottom valves !!!

The result may look like a volcano explosion. The reason is that the fermenting must is supersaturated with CO_2 gas (especially during cold fermentation of white wine), and if some perturbation happens in the liquid (like a wave from the incoming must from the bottom), it may release the extra dissolved CO_2 all at once, which would look like an explosion. If addition of must is necessary to transfer into a fermenting tank, it should be done slowly and only from the *top* of the tank.

It should also be noted that any stage of fermentation can be checked microscopically, by counting the yeast cells (alive and dead) and by watching the normal budding of the yeast cells. It is also necessary to follow temperature changes during fermentation and prevent any drastic changes due to failure of the cooling system. Toward the end of fermentation, analysis of alcohol, residual sugar, volatile acidity, total acidity, pH, and sulfur-dioxide, give the winemaker the basic information needed on the processed wine.

2. Stuck Fermentation

Almost any winemaker has had the experience of stuck fermentation at any stage of its progress, from the beginning and up towards the end, when a few Brix are still unfermented. Knowing the possible reasons that may limit or inhibit yeast growth at the different phases of the fermentation process will help to prevent sluggish fermentation from occurring, or in case it happens, to revitalize it. The reasons might be:

Lack of oxygen: Although fermentation is anaerobic, the use of oxygen at the second phase of fermentation (exponential growth) is of crucial necessity. It is needed for utilizing the steroid *ergosterol* and unsaturated fatty *oleic acid* ($C_{18}H_{34}O_2$), for cell membrane building. Without it the yeast will not be able to multiply and increase their biomass needed to complete the fermentation. In a study where air and nitrogen gas were flushed through experimental samples, the nitrogen-flushed fermentation was extremely sluggish and stuck. On aeration the stuck fermentation was restarted, and kept on until completion. Yeast population ratios between the aerated and nitrogen-flushed samples were over 20-fold higher. Commercial fresh dry yeast is added at about 20 gr/HL (about 2Lb to 1000 gallons) of must. At such concentration there are a few million yeast cells/ml in the freshly inoculated must. At maximum activity, the yeast cell count stabilizes at about 100 million/ml, (at the end of phase-II). During the exponential phase-II the yeast have to multiply (5 – 6) times ($2^5 – 2^6$) in order to increase the population from a few million/ml to 100 million/ml. If there is not enough oxygen in the must, the growth will suffer and the fermentation will get stuck. In this case, in order to aerate the must, pumping about a quarter of its volume over the top of the upper surface of the open tank will do the job. A good practice in white must fermentation is to rack the fermenting must two days after inoculation into another tank. This racking will aerate the must right at the end of the lag period and also homogenize it (yeast culture and nutrient).

Lack of nutrient: Especially nitrogen in the ammonia form. Some winemakers regularly add DAP (diammonium-phosphate) $(NH_4)_2HPO_4$ to the must before starting fermentation. It is generally less needed in red wine fermentation, but it is recommended in white clear must after lees racking. The added quantity is (10 – 20) gr/HL of DAP to prevent fermentation problems such as stuck fermentation or "stinky fermentation" caused by lack of nitrogen.

To complete the nutrient deficiency (rare amino-acids, vitamins, and minerals), there are also commercial yeast nutrients, sometimes called "yeast extract," which contain mixtures of all the necessary ingredients at the right proportions. In cases of stuck fermentation is it advisable to use such extract at quantities of about (20 – 25) gr/HL.

Temperature: The effect of temperature on fermentation rate is normal as in other biological enzymatic reactions. It increases the activity rate within the middle range of activity, and decreases it at the extreme edges, until causing it to become stuck at certain temperature values. This means that when the temperature gets higher beyond the activity range it inhibits fermentation, and temperature that is lower beyond that range also inhibits it.

In white wine, which is usually fermented at (8 – 12) $^{\circ}$C, the lag period may sometimes be very long. Acclimation of the yeast to low pH, high osmotic pressure of sugar and very low temperature, might be difficult, causing delay in adaptation. If the yeast is viable, increasing the temperature by a few degrees may accelerate its acclimation and open up the second phase of fermentation. At this stage, fermentation heat is released, and as soon as the fermentation is on its way, re-cooling back is necessary. To avoid this problem, it is better to inoculate the must when the temperature is higher than the desired fermenting temperature, for example to inoculate at 17°C and only after good signs of fermentation by reduction of (1 – 2) Brix and CO_2 bubbling, to reduce the temperature to the desired one. In a case where the cold fermentation starts well but then becomes stuck, it may work well to elevate the temperature a few degrees and add a new batch of yeast. After fermentation resumes, the temperature can be lowered again.

At the higher edge of the scale, when fermenting red wine without cooling, or when the cooling system fails, the temperature might rise up to (35 – 40) $^{\circ}$C. Above 35°C the high temperature and the alcohol content create a deadly environment for yeast. The yeast has difficulty in maintaining the vital enzymatic reactions, so in red wine fermentation, especially in large tanks, careful watching of the temperature is advised. If the temperature gets too high and fermentation stops, cooling the must to about (25 – 30) $^{\circ}$C and adding a new batch of yeast (after gradual acclimation in the unfinished wine) may restart the fermentation and bring it to completion.

High sugar concentration: When the must has very high sugar content (above 25B$^{\circ}$), the physical and chemical conditions promise some problems to come during fermentation. First, the osmotic pressure of such a solution is high, which makes it hard for the yeast cells to survive and to multiply. Therefore, some measures must be taken to enhance the chances of having straight fermentation without being stuck. Without these measures, the wine might be infected by undesired yeast species or bacteria, causing undesirable off-flavor. Also, high residual sugar might be left over after fermentation, leaving the wine vulnerable to further instability, and to be not completely dry. The measures are:

(1) To select a yeast strain with high sugar and high alcohol tolerance so that

they can ferment the sugar to dryness up to (15 – 16) % of alcohol.

(2) To add a high dose of yeast at the start, namely (250 – 300) gram/ton of must, or even more, depending on how high the B° is. For late harvest grapes with very high sugar concentration of the order of (30 – 35) B°, 500 gram/ton will be better.

(3) To add a high dose of SO_2 to prevent the forecast problems due to the delay in finishing the fermentation. (50 – 70) ppm of SO_2 will be needed.

(4) To select two (or more) yeast strains in combination, to make sure that at least one of them will take the task of completing the fermentation when the other fails to do so. Select with care, based on their characteristics to match the pair combination. For example, use Bordeaux-red yeast strain for Cabernet Sauvignon grapes to get good-flavored wine; however, this strain is sensitive to killer yeast problems and the fermentation might be stuck quite frequently. So you may use as the second strain a strong fermenting yeast which is tolerant to high alcohol in the wine and can bring the fermentation to dryness. You may try to add the second yeast not in the beginning of fermentation, but later, say at (5 – 10) B° reduction. The conditions will still enable the new young yeast batch to acclimate easily.

(5) To add DAP at (100 – 200) mg/L range to the must before fermentation, to avoid nutrient deficiency. In regular fermentation it is not necessary, but in a problematic one it may help to avoid difficulties.

(6) And lastly, in order to get better balanced and reasonable wine, avoid high temperatures during fermentation, to reduce the VA production, and be aware of the pH. If it is too high (over 3.8), consider reducing it by acid correction before fermentation.

All these measures might help to finish the fermentation safely to dryness.

Fatty acids: It was well recognized that fatty acids, mainly *octanoic-acid* {CH_3-$(CH_2)_6$–COOH} and *decanoic-acid* {CH_3-$(CH_2)_8$–COOH}, are the major responsible factors for the inhibition. Intermediate size fatty acids with (6 – 12) carbons, namely, from *hexanoic* to *dodecanoic* acids, are synthesized during fermentation and accumulate as self-toxic agents to yeast growth. They are not by-products of the fermentation per se, but are "debris" parts left over unused during synthesis of long chain lipid acids, needed for cell membranes. Being hydrophobic, the fatty acids can enter the yeast cell's membrane and interfere with the transport systems between the cell and its medium. The concentration range of *octanoic* and *decanoic* acids found in wine at the end of alcoholic fermentation is about (2 – 10) mg/L. Towards the end, their accumulation can stop the fermentation. By elimination of all other known factors, it appears that this might be the most probable cause of stuck fermentation. For a long time, it has been known that adding activated carbon can stimulate, and sometimes reactivate fermentation that has become stuck in its advanced state. In fact, activated carbon adsorbs the fatty acids from the must, removing these toxic agents, which then enables the fermentation to proceed. Dead yeast cells, or "yeast-ghost" as they are commercially called, can do the same thing very efficiently. The potential of "yeast-ghost" to restart a stuck fermentation has been verified in experiments. The "yeast-ghost" can be added prior to fermentation, or after a few days.

Unviable yeast: If fermentation does not start, or continues to be stuck after all the above measures have been tried, a new yeast starter should be introduced. And when you have to restart your stuck fermenting wine, you should do it with some guidelines:

(1) Select the strongest strain available, taking this character as the main choice and preferably not using the same yeast strain you used initially for that fermentation.

(2) Use a much higher dose of yeast to restart, namely (50 – 100) gram/HL of wine.

(3) To start such a stuck fermentation, the new starter must be acclimated to the alcohol and other by-products, which are contained in the environment of the stuck must. In such cases the starter should be handled as follows:

After re-hydration with water for ten minutes, add to it about a third of its volume from the stuck must and measure the Brix of this culture. Allow it to ferment until the Brix decreases about (1 – 2) units and then add another batch of the stuck must. When it continues to ferment (by Brix measurement), add it to the entire must.

Killer yeast: Mentioned earlier in this chapter.

3. Fermentation By-Products

Besides ethyl alcohol, which is the major product of alcoholic fermentation, there are many other by-products found in wine as a result of the fermentation process. Some of them are desired and contribute to the wine quality, but some others are harmful and may cause damage to the smell and taste of the wine. We will discuss here the major substances produced during fermentation. Some of them are direct by-products of the fermentation path, while others are formed as a result of other reactions taking place in the must medium during fermentation.

Acids

The following acids exist in wine, although their origin is not in grapes. Their accumulating content contributes, together with the grape acids, to the total acidity and the pH in the wine.

Succinic Acid : $COOH - CH_2 - CH_2 - COOH$

This acid is formed during fermentation as direct by-product in all alcoholic fermented beverages and contributes its share to the total acidity. The range of concentration found in wine is (0.5 – 1.5) gr/L. The acid is very stable and does not change during aging.

Lactic Acid:

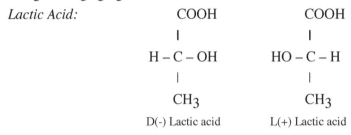

D(-) Lactic acid   L(+) Lactic acid

Lactic acid is formed in wine by two pathways: as a by-product of yeast fermentation, and as the main product in the ML-fermentation of malic acid. D(-) is the absolute S configuration, rotating polarized light to the left. L(+) is the R absolute configuration, with right rotation.

Yeast fermentation produces a racemic lactic acid (a mixture of D and L, called also DL-lactic acid), with some preference to the form D(-) lactic acid. The concentration contribution from this pathway is about (0.2 – 0.4) gr/L. On the other hand, malolactic-bacteria produce *exclusively* L(+) lactic acid from L(-) malic acid, the original malic acid in grape juice. (More details are presented ahead in this chapter.)

Acetic Acid: $CH_3 - COOH$

Acetic acid is the main volatile acid in wine, which may also contain other volatile acids such as formic, propionic and butyric acids (all in very small concentrations). The term VA (volatile acidity) is used to define all these acids as one group in which acetic acid is the dominant one. Acetic acid is formed as a by-product of alcoholic fermentation, as a result of the side reaction of acetaldehyde oxidation to ethanol. The normal range of acetaldehyde formation is about (300 – 500) mg/L. At this level it is not noticeable on the palate and has no effect on wine quality. In Botrytis grapes (and wines) the usual concentration is much higher and accepted as part of this style of wine. Some great Sauternes and California late harvest wines contain typical VA concentrations in the $(1.0 - 2.0)$ gr/L range. Acetic acid can also be formed by acetobacter spoilage bacteria, in an aerobic condition. The bacteria oxidize the wine's ethanol to acetic acid at concentrations which depend on the air exposure and time. In extreme cases it may convert the whole alcohol content into vinegar. Above $(1.0 - 1.2)$ gr/L VA is noticeable and it depreciates wine quality. The legal limits of VA content in the US are 1.2 gr/L and 1.4 gr/L for white and red wines respectively.

Alcohols

Methanol: CH_3- OH

Methanol is not a direct product of fermentation. The source for methanol found in alcoholic fermentation of fruit products is the fruit pectin which is hydrolyzed by *methylesterase* enzymes. Pectin is a co-polymer of galacturonic-acid and its methyl ester. Hydrolysis of the methyl ester yields methanol. Hydrolytic enzymes exist naturally in the must. Addition of pectolytic enzyme to the must in order to facilitate must clarification also increases methanol content. Because pectin is found more in the skin than in the juice, white wines contain much less methanol than red wines, which are fermented with the skin.

A wide survey on reported methanol content in wines around the world showed that the average concentration in white wines is 60 mg/L in $(40 - 120)$ mg/L range, and 150 mg/L in red wines in $(120 - 250)$ mg/L range. In extreme cases the content may be as high as several hundred mg/L. The temperature of fermentation does not affect the methanol content in wine; however, as stated above, pectin treatment and prolonged skin contact do.

Although methanol is toxic, its concentrations in wines do not constitute any risk. The lethal toxicity dose (Ld_{50}) is about 340 mg/Kg of body weight. Taking into account methanol content in wine, say about 100 mg/L, and using the toxicity value of 340 mg/Kg, a person whose weight is 140 lb. ($\sim$70 Kg) has to drink about 240 liters of wine (!) in order to be intoxicated. Methanol is metabolized in the body like ethanol (although at a slower rate). The legal limit of methanol in wine in the U.S. is 1000 mg/L.

Higher Alcohols (Fusel Oils)

These are the homologous series of saturated alcohols, starting with propanol.

Their presence in fermented products has been recognized for a long time, and some mechanisms have been proposed for their production. The major sources for higher alcohols are amino acids, which by a sequence process of trans-amination, decarboxylation, and reduction, are transformed into alcohols as follows: Amino acid $R - CH(NH_2) - COOH$ losing its amino group which is then de-carboxylated to aldehyde (losing one of its carbons), and then reduced to alcohol. For example:

$$(CH_3)_2 - CH - CH_2 - CH(NH_2) - COOH \longrightarrow (CH_3)_2 - CH - CH_2 - CO - COOH \longrightarrow$$
$$\text{leucine} \qquad\qquad\qquad \text{α-keto-isocaproic acid}$$

$$\longrightarrow (CH_3)_2 - CH - CH_2 - CHO \longrightarrow (CH_3)_2 - CH - CH_2 - CH_2OH$$
$$\text{isovaleraldehyde} \qquad\qquad \text{isoamyl alcohol}$$

The final result is that an *amino acid* with n-carbons is transformed into an *alcohol* with (n -1) carbons. When there is a shortage of specific amino acids in the must during fermentation, these trans-amination reactions take place in order to supply the needs for fermentation, and thus higher concentrations of higher alcohols are formed as by-products of the amino acids' decomposition.

The major high alcohols found in wine are:

n-propanol: $\qquad\qquad CH_3 - CH_2 - CH_2OH$

iso propanol: $\qquad\quad (CH_3)_2 - CHOH$

n-butanol : $\qquad\qquad CH_3 - CH_2 - CH_2 - CH_2OH$

iso butanol: $\qquad\qquad (CH_3)_2 - CH - CH_2OH$

n-amyl alcohol: $\qquad CH_3 - CH_2 - CH_2 - CH_2 - CH_2OH$

3-methylbutanol: $\qquad (CH_3)_2 - CH - CH_2 - CH_2OH$

2-methylb nol: $\qquad CH_3 - CH_2 - CH(CH_3) - CH_2OH$

n-hexanol: $\qquad\qquad CH_3 - CH_2 - CH_2 - CH_2 - CH_2 - CH_2OH$

2-phenylethanol: $\qquad (C_6H_5) - CH_2 - CH_2OH$

Their total concentration range in wine is between (100 – 500) mg/L.

2,3-Butandiol: $\qquad\quad CH_3 - CH(OH) - CH(OH) - CH_3$

This alcohol is the major *di-alcohol* found in wine. Its concentration was found to range between (0.3 – 1.8) gr/L with an average value of (0.5 – 0.8) g/L.

Although we are not yet completely sure how it is formed, it is a by-product of alcoholic fermentation, probably from pyruvic acid, or by reduction of acetoin ($CH_3 - CH(OH) - CO - CH_3$).

glycerol: $\qquad\qquad\qquad CH_2(OH) - CH(OH) - CH_2(OH)$

Glycerol is the major *tri-alcohol* found in wine. It is formed as a byproduct of sugar fermentation, mainly at the beginning of the fermentation, through side-chain reactions which end up as glycerol. Glycerol content is related to the amount of ethanol formed, to the temperature of fermentation, and to the yeast strain. It is higher at higher fermentation temperatures, so its content is higher in red wines than in whites. In a study of California wines, an average value of 4.5 gr/L was found in white and rosé wines, and 6.5 gr/L in reds. The content of glycerol reported in a wide variety of wines around the world is in the range of (4 – 12) gr/L. A general conclusion can be drawn that glycerol content in wine is between the ranges of (6 – 12) % of the ethanol content. The lower range represents white wines, and the higher range red ones. In late harvest or botrytis wines, the glycerol concentration is significantly higher than in regular table wines. Concentrations of (15 – 25) gr/L and above have been reported. In this case, glycerol was synthesized by the *Botrytis-cinerea* fungi, and became more concentrated in the dried berries. Some amount of glycerol exists also in healthy grapes (up to 0.5 gr/L). Because glycerol is very viscous and sweet (about 70% of glucose sweetness), it is assumed that it has some contribution to the wine body and mouth feel.

erythritol: $CH_2(OH) – CH(OH) – CH(OH) – CH_2(OH)$

Erythritol is a *tetra-alcohol* formed from the 4-carbon sugar *erythrose* $CHO – CH(OH) – CH(OH) – CH_2OH$. Note that erythritol is a meso-compound (having internal symmetry). Its sweetness is about twice that of sucrose. Erythritol is formed during fermentation, and it was found to depend very much on the yeast strain. Its concentration in table wines, which were fermented by *Saccharomyces-cerevisiae* yeast, is in the (30 – 100) mg/L range. Other yeast species such as *Kloeckera-apiculata* and other "wild" yeasts can produce several hundred mg/L of erythritol. In *Botrytis-cinerea* infected grapes the concentration of erythritol is in the (50 – 600) mg/L range.

arabitol: $CH_2(OH) – CH(OH) – CH(OH) – CH(OH) – CH_2(OH)$

Arabitol is a *penta-alcohol*, which was reported to be found in wine in the range of (10 – 60) mg/L in white wines and (30 – 110) mg/L in reds. In high quality German wines infected by *Botrytis-cinerea* the concentrations found were much higher, in the hundreds of mg/L range, and in several cases up to 2300 mg/L.

mannitol: $CH_2(OH) – [CH(OH)]_4 – CH_2(OH)$

Mannitol is a *hexa-alcohol* formed from fructose during fermentation and is found in wines in the range of (100 – 400) mg/L. Late harvest German wines showed very high concentrations of mannitol from hundreds of mg/L up to several grams/L. In this case it is related to the delaying of the harvest. (Mannitol can also be formed by lactic acid bacteria, which is always followed by acetic acid formation. In such cases the wine is considered spoiled because of the acetic nose.)

Sorbitol is another hexa-alcohol, an isomer of mannitol. It is found in wines in a concentration range of (50 – 200) mg/L, and in late harvest German wines, as much as few hundred mg/L up to 1000 mg/L. As with other polyalcohols it is associated with the noble rot grapes and wines.

Aldehydes and Ketones

Aldehydes and ketones are formed during alcoholic fermentation. The main representatives that are found in wine are *acetaldehyde*, *acetoin* and *di- acetyl*.

acetaldehyde: $CH_3 - CHO$

Acetaldehyde is the last major component in the fermentation process chain from sugar to acetaldehyde, which normally is then enzymatically reduced to the final product, ethanol. But some of it may not be reduced, and it remains in the wine in a concentration range of (20 – 200) mg/L, with an average content in table wines of about 70 mg/L. In aged wines its concentration may increase by a slow non-enzymatic oxidation of ethanol to acetaldehyde:

$$CH_3 - CH_2 - OH \overset{(o)}{-----} > CH_3 - CHO$$

At bottling, when wine is highly exposed to air, an extra amount of acetaldehyde may be formed which gives the wine some flat taste. This phenomenon is sometimes called "bottle-sickness," and it disappears with time. Enough free SO_2 before bottling will minimize this effect.

Flor-sherry yeast under aerobic conditions oxidizes ethanol to acetaldehyde (and some other aldehydes such as propionaldehyde, isobutyraldehyde and iso-valeraldehyde). These sources of aldehydes, including acetaldehyde, are not by-products of fermentation and have no connection to fermentation at all. In *sherry wines* the average concentration of acetaldehyde is very much above 200 mg/L. Acetaldehyde has a very distinct nose, which characterizes *sherry wines*, but it is not thought to contribute to the quality of *table wines* when it is noticeable by its smell.

acetoin: $CH_3 - CH(OH) - CO - CH_3$

This ketone is formed as a by-product of fermentation. During fermentation the acetoin concentration reaches a maximum value of about 100 mg/L, which later decreases (probably by reduction to 2,3-butanediol), to (5 – 20) mg/L.

In dessert port-type wines where the fermentation is halted in the middle by alcohol addition, the acetoin concentration is found to be higher than average, in the (20 – 50) mg/L range.

diacetyl: $CH_3 - CO - CO - CH_3$

This di-ketone is also a by-product of alcoholic fermentation. It is found in very low concentrations. Red table wines which have undergone malolactic-fermentation have an average of 2.8 mg/L, and those which have not, have a lower average range of 1.3 mg/L. Thus, malolactic-fermentation contributes *additional* diacetyl to the regular post-fermentation content. The detection threshold of diacetyl is about 1 mg/L. It is estimated that at (2 – 4) mg/L range, diacetyl contributes to the quality of red table wine by its buttery aroma, making it more complex. Above this range the quality is reduced, and diacetyl is recognized as an off-flavor nose.

Esters

Esters are the product of the reaction between alcohols and acids. In wine they are classified into two categories: *neutral-esters* and *acid-esters*. In *neutral-esters* the acid part of the esters comes as a by-product of fermentation and they are formed through an enzymatic process. They include acetate, butyrate, hexanoate, and other esters. In *acid-esters* the acids are mainly the original grape acids (tartaric and malic), formed by a regular chemical esterification between alcohols and the wine acids. In wine we find esters from three sources:

(1) Originally produced by the grapes, in very small amounts.

(2) Produced by enzymatic esterification during fermentation, from by-product acids of the fermentation process (neutral-esters).

(3) Produced by very slow chemical esterification during aging of the wine, from tartaric and malic acids (acid-esters).

The neutral-esters, which are volatile aroma compounds, are also called *volatile esters*. Our interest here is in this group, which is an important factor in wine nose and quality. The concentrations of the volatile esters in wine are low, and their organoleptic perception is described as a fruity-type aroma, except for *ethyl acetate*, which has the highest concentration in wines, and its contribution to wine quality is considered negative.

The acid-esters on the other hand are not volatile, without specific organoleptic character. They are called *non-volatile esters*.

ethyl acetate: $CH_3 - CH_2 -O- CO - CH_3$

It is always accompanied by acetic acid (its acid precursor), and in fact it contributes the distinctive vinegar essence to wines when its concentration is higher than (160 – 180) mg/L, in comparison to acetic acid which starts to be noticeable (only by taste as a hard sourness) at above 750 mg/L.

The actual concentration of ethyl acetate found in wines as a by-product of fermentation is about (30 – 70) mg/L. On the other hand, wines spoiled by *acetic acid bacteria* at aerobic conditions have much higher concentrations of ethyl acetate, where the main product of this spoilage is acetic acid.

Higher Volatile Esters

Beside the above negative ester (ethyl-acetate), young wines contain higher volatile esters, which have great significance, mainly in young white wines in relation to their fresh fruity aroma. They are also very important components in brandy, where their concentrations are raised in the distilled brandy. When young wine is aged, its volatile esters are gradually decreased by hydrolysis, losing their fresh fruity aroma. Some of the most dominant esters are:

Isoamyl Acetate: $(CH_3)_2 - CH - CH_2 - CH_2 -O- CO- CH_3$
Found in concentration range of (2 – 6) mg/L, with odor threshold of about (0.16 – 0.23) mg/L.

Ethyl Octanoate: $CH_3 - (CH_2)_6- CO-O- CH_2 - CH_3$
concentration range of (0.5 – 2) mg/L with odor threshold of (0.2 – 0.6) mg/L.

Ethyl Hexanoate: $CH_3 - (CH_2)_4- CO-O- CH_2 - CH_3$
concentration range of (0.2 – 1.5) mg/L with odor threshold of 0.08 mg/L.

Ethyl Decanoate: $CH_3 - (CH_2)_8 - CO-O- CH_2 - CH_3$
concentration range of (0.1 – 1) mg/L with odor threshold of (0.5 – 1) mg/L.
This is only a partial list of about 300 esters which have been found and listed in wines.

4. Residual Sugar
Classification

The term *residual sugar* refers to any significant concentration of sugar that is contained in wines. It may vary from (1 – 3) gr/L in very dry table wine up to

(50 – 150) gr/L in port style wines or late harvest botrytis dessert wines. At any fermentation, when going smoothly, there is some residual sugar left over. These residual sugars contain unfermentable pentoses (*arabinose* and *rhamnose*) and some fermentable hexoses (*fructose* and *glucose*). If the fermentation is stopped before all fermentable sugars have been consumed, more fructose will be found than glucose, because the glucose rate of fermentation is slightly higher than that of fructose. Also, besides sugar, there are other sweet components in wine, such as glycerol, at a concentration range of (4 – 12) gr/L with about 70% of the sweetness of glucose. Ethyl alcohol has also a slightly sweet taste, so even when the fermentation has come to "complete" dryness, some people experience a sweet taste in dry table wine, especially whites, but also in reds. In good quality wines containing sugar, no addition of cane-sugar is accepted. The sweetness should come only from the natural grape sugar, either by sugar left over after fermentation or by addition of grape juice or grape concentrate to a dry wine to make it sweet. This is the common practice; it is legal and well recognized.

There are many types of wines containing sugar in the market. They can be classified into four major categories:

Off-dry table wines – with about (5 – 30) gr/L sugar. Certain table wines, especially if they have fruity-flowery and fresh character, are better balanced with some sweetness. In Muscat-related varieties (Muscat of Alexandria, Muscat Canelli, Riesling, Gewurztraminer, Symphony), it is almost essential to leave some residual sugar in order to match with the very spicy and powerful perfumed aroma of these varieties. In certain cases it is necessary to mask the bitter aftertaste that sometimes characterizes these varieties. Also, residual sweetness is quite common in the "blush" wines ("blanc de noir").

Semi-sweet wines – mainly whites with a higher sugar content of about (30 – 50) gr/L. The low-medium sugar content is necessary to balance the high acidity content of these wines. It is typical of grapes grown in cool climates.

Sweet dessert wines – such as Port, Sherry (Oloroso), Madeira, Marsala, Vermouth, and others. Each wine in this category has its special character, depending on its unique version of production. The residual sugar content in these wines is usually between (50 – 100) gr/L.

Late harvest wines – with two main styles:

1. German late harvest style – (Auslese, Beerenauslese, Eiswein) which are made of overripe grapes where the sugar level gets higher than the regular level. In certain cases where the grapes become shriveled and almost dry, the sugar level is very high. The most popular varieties in this version are White Riesling and Gewurztraminer. Hungarian Tokay can also be classified here.

2. *Sauterne late-harvest (Botrytis) style* – wines which are made of late harvest grapes infected by *Botrytis-Cinerea* mold ("noble rot"). The infection, which usually causes the grapes to become "rotten" and undesirable as table wine, may in certain weather conditions (high humidity at infection time, and dry and warm in the following days) cause them to develop in such a way that the berries lose their water without becoming rotten. The sugar level (with the dry extract) gets very high, rising to about (30 – 50) Brix. The Botrytis not only concentrates the dry extract in the berries, but also gives some very special aroma characteristics to that style of wines. The wine made from such grapes ends up with about (50 – 150) gr/L of residual sugar. The most common varieties for this type of wines are Sauvignon Blanc, Semillon, White Riesling and Gewurztraminer. The production of these wines is quite different from that of table wine. The right time to pick the grapes is most crucial. If picked too early, the sugar will not be concentrated enough, and if picked too late, the grapes might be totally rotten or dry. Several pickings may be necessary according to the bunch "noble rot" development, and even individual berry picking is practiced in some places.

The pressing of the shriveled berries (whole clusters, no destemming/crushing), needs high pressure for a very long time, and many cycles of pressing. The juice contains a very high percentage of solid lees (which will be difficult to clear up), and a high content of aldehydes, volatile acids, and glycerol. No settling is done, generally, and the usual practice is to add about 100 ppm SO_2 before fermentation, most of which will probably be fixed by the aldehydes. Diammonium-phosphate (DAP) is also recommended (200 ppm), to ease the fermentation. The inoculation is best done with yeast culture rather than with dry yeast, in order to acclimate the culture gradually to the very high sugar level of the must. It is also preferable to start the fermentation at a moderate temperature (20 – 25) °C, and only after it has started well, to reduce it to the desired lower temperature. The fermentation may take a very long time to finish, until it stops at a certain sugar level. It can then be aged in oak barrels for 6 – 18 months before bottling. In certain places the must itself is fermented in oak barrels. Addition of SO_2 is necessary to maintain about (20 – 30) ppm of free SO_2 during all the processing stages up to bottling. Because of the initial high SO_2 and the constant addition during vinification, the total SO_2 in late harvest wines is usually considerably higher than in regular table wines. A common level of SO_2 in late harvest wines is (200 – 250) ppm.

Stopping Fermentation and Preservation

There are several techniques to stop the fermentation before dryness and to pre-serve it from re-fermenting. Stopping fermentation is the "natural" method used to leave a certain amount of residual sugar in the finished wine. It can be done in any of the sweet categories: fully sweet, semi-sweet dessert wines or off-dry table wines.

Deep cooling: Most suitable for premium table wines. If this technique has been chosen, right from the beginning the winemaker should make the yeast's condition somewhat uncomfortable. Under some stress the yeast will be weak enough to be easily controlled. No nutrition should be added before fermentation (lees settling is obvious and, using bentonite prior to fermentation is also possible). Yeast inocula-tion can be done with a lower quantity of the regular amount, and low fermentation temperature (10 – 12) °C is the main regulation factor.

Although any yeast strain can be used, the yeast strain that is easiest to stop by cold-shock is *Epernay-2*. This strain is also good for fruity and highly aromatic white wines. Under all these conditions, the fermentation may last longer than usu-al, up to (4 – 6) weeks. Toward the end of the fermentation, the alcohol and sugar analysis should be made and, at the desired sugar concentration, the temperature should be drastically lowered under 4°C (39°F). The fermentation will then termi-nate. If the fermentation is stopped while the yeast is very active and the rate of the fermentation is still high, there may be some delay before the final termination. If a strong must chiller is not available to immediately cool the must, it will take some time for the temperature to drop down just by reducing the temperature control. In such a case, deciding when to stop the fermentation should be based on extrapolat-ing the Brix reading above the desired residual sugar level. For such extrapolation, a daily record of the Brix level is necessary. In about a week or two, the wine has to be racked off from the yeast lees, sulfited, bentonited, and cold-stabilized. After cold stabilization, the wine should be filtered and aged in a tank until bottling at low temperatures not exceeding (5 – 8) °C.

Natural stopping: In some cases, just by imposing stress on the yeast, the fer-mentation may stop by itself. The stopping is unpredictable, with no control over the desired residual sugar level in the wine. In late harvest must, the conditions are such that the fermentation is always under stress. The nutrition is usually too low, and the high sugar concentration is an inhibition factor for the fermentation, especially when the alcohol becomes high enough. So, in most cases, late harvest must will stop its fermentation naturally.

Alcohol fortification: Another technique to stop fermentation and leave residual sugar is by addition of alcohol to the fermenting must up to a final concentration of about 17% (v/v), where the yeast cannot tolerate any more and dies. The addition of alcohol is done at the desired sugar concentration, as in dessert wine production like Port, Sherry, Madeira, and others. Because of the sugar content, lower alcohol concentration—less than 17%—is needed in order to inhibit the fermentation. The combination of high alcohol and sugar concentration will preserve the wine from re-fermenting during the rest of the wine processing and in the bottle. (For more details on how much alcohol to add, the reader is referred to the *Concept of Wine Chemistry* book, chapter VIII, section D). After the addition of alcohol, the wine is well mixed, racked from the yeast lees, and cold-stabilized.

Sweetening dry wine: There is a different technique to introduce residual sugar into wine that is less troublesome, and less caution is needed. This is most suitable for regular, less expensive, off-dry table wines. In this technique the wine is fermented to dryness, racked off from its yeast, sulfited, and bentonited. After bentonite racking and filtering, a calculated amount of sugar is added as grape juice concentrate, to bring the wine to the desired sugar level. Stabilizing and cellar aging are carried on as with any regular wine, but until bottling the wine should be kept at a low temperature of less than (5 – 8) °C to prevent re-fermentation. Another way to sweeten dry wine is by adding "sweet reserve" (unfermented grape juice) which has been kept in the cellar since harvest. The "sweet reserve" is preserved by lees settling to clear must, then sulfiting up to (400 – 500) ppm of SO_2 and storing at a cool temperature close to 0°C.

Less care is needed if the sugar addition is made just before bottling, where the addition is followed immediately by micro-filtration. The only risk with such late sugar addition is that it may change slightly the wine's stability in the bottle. However, if the wine is to be consumed young, this risk is minimal.

Sugar Preservation in the Bottle

Any fermentable sugar is a major microbiological instability factor in bottled wine, and special attention has to be paid to maintain it. This section deals with some ways to preserve wine from re-fermentation.

Yeast Inhibitors: The most common agent to inhibit fermentation is *sorbic acid* or its salts. Before bottling, about (200 – 250) ppm of potassium-sorbate can be added. The sorbate has an inhibitory effect on yeast growth, and no less than 200 ppm should be used in order to be effective. In complementary action with the sorbate, addition of (30 – 40) ppm of free sulfur-dioxide is highly recommended. The sorbate has almost no offensive taste or odor at these levels, especially in sweet wines. Its taste threshold is about 150 ppm, and its legal limit in the United States is 1,000 ppm. The main objection to the use of sorbate is the "geranium-like" odor which may develop by lactic bacteria from sorbic acid, to produce *2-ethoxyhexa-3,5-diene*. To some people this compound may be pleasant, to others unpleasant. The SO_2 presence with the sorbate is a good assurance that ML-bacteria will not be active to carry on that reaction. The sorbate is widely used in sweet wines that have not been fortified or pasteurized.

Pasteurization: This is a very old and very efficient technique to prevent re-fermentation of sweet or semi-sweet wines. It is not recommended for quality table wine because the heat will destroy its freshness and fruity aroma. It also may contribute a caramel taste to the wine. The lowest damage is caused by flash pasteurization at about 80°C for a few seconds, followed immediately by cooling. After the pasteurization, the wine may become cloudy because of protein precipitation, so fining with bentonite and filtration are necessary. However, by doing so, the wine may become un-sterile again from the tanks and the equipment used, so another pre-bottling pasteurization is needed. To avoid it, the wine can be bottled directly after the bentonite racking through pad filtering and micro-filtering to the bottling line.

Sterile Filtration: This technique is based on mechanically removing the yeast from the wine, and keeping it under conditions that make fermentation very difficult to restart (low nutrient, alcohol, SO_2, and almost null yeast population). Up to bottling, when sterile filtering is done, the wine has to be checked regularly for signs of re-fermentation (CO_2 bubbles, microscopic test). At bottling, sterile filtration (0.45μ) prevents any wine spoilage microorganism from getting through. In the modern technology of winemaking, this is the most commonly used and effective technique to prevent re-fermentation in the bottle.

5. Heat of Fermentation

Heat and Cooling

From the yeast point of view, fermentation is carried on in order to get the energy needed for living. Analysis of the energy balance of the fermentation process is of great importance in understanding the temperature changes that occur during fermentation. This can be very useful in designing suitable cooling systems for controlling fermentation rate and temperature. In our case, part of the energy released by the fermentation reaction (converting sugar into alcohol and CO_2) is consumed by the yeast for their living and reproduction. The extra energy (waste) which is not used by the yeast is released as heat which causes the must temperature to rise.

The fermentation reaction can be written to include the energy ∆E released as a part of the changes that occur in the reaction:

$$C_6H_{12}O_6 \text{ ----------> } 2CH_3 - CH_2 - OH + 2CO_2 + ∆E$$

| MW = 180 | MW = 46 | MW = 44 |

∆E = 56.4 Kcal/mole.

Out of the total theoretical energy ∆E = 56.4 Kcal/mole, the yeasts *utilize for their well-being only 32.4 Kcal/mole.* The rest (waste), **24 Kcal/mole**, is released as heat. This calculation is based on the pure reaction of converting glucose into ethanol and CO_2. In real fermentation some other factors play a part, such as production of other by-products (glycerol, succinic acid) and compounds which are used as yeast biomass. Taking most of these factors into account, including their relative molecular ratio, reveals that the actual heat waste during alcoholic fermentation that is reported in the literature (theoretical and experimental) is slightly lower than the pure theoretical one, and it is between the range of **(22.5 – 24) Kcal/mole**.

(Before we move to a practical calculation of heat, temperatures, and cooling capacity, it is worthwhile to note that if the yeast could ferment sugar *aerobically* to CO_2 and water, i.e., be able to use the high energy content in ethanol, 632 Kcal/mole of sugar could be gained, in comparison to 56.4 Kcal/mole. And assuming the same yield as before (57%), 360 Kcal/mole could be used for well-being instead of 32.4 Kcal/mole. That is about ten times more than by the *anaerobic* process.)

Let us calculate the amount of heat that is actually released in the fermentation process. We should know that the 24 Kcal/mole of sugar is dissipated in two ways: by heat, and by the mechanical work that the CO_2 is doing by evolving into the atmosphere. To make it general and later on of practical use, the calculation will be done for 1Brix (10 gram/L) of must sugar. We will calculate the heat released by **one Brix**, and then we will be able to use it for any must with its specific Brix.

An average of 10% of the solid Brix, represent non-fermenting components in the must (acids, phenols, minerals, non-fermented hydrocarbons, and others) ===> 90% of 1^oB = 9 gram sugar/L.

a. We show above that 1 mole sugar (180 gram) will produce ----> 24 Kcal ; 9 gram will produce: 24 x 9/180 = **1.2 Kcal**.

b. To check how much energy is wasted by the CO_2 work, let us make the following calculations:

One mole of sugar (180 gram) will release 2 moles of CO_2;

9 gram will release: 2 x 9/180 = 0.1 mole CO_2.

The volume of 1 mole gas at standard conditions (1 atm. pressure and 0^oC) is known to occupy 22.4 L. Correction for 20^oC makes it 24.0 L;

0.1 mole of CO_2 will occupy 2.4 L at 20^oC.

c. The CO_2 evolution is done against atmospheric pressure which is $1kg/cm^2$. The mechanical work which is done by the gas evolving into the atmosphere is the *pressure* x *volume*, namely, $1Kg/cm^2$ x $2400cm^3$ = 2400 Kg x cm ;

The conversion factor to joule is: 1Joule = 1Newton x m

==> 2400 kg x cm = $2400 x 10 x 10^{-2}$ = 240 Joule.

The conversion factor between joule and Kcal is: 1Kjoule ==> 0.24xKcal. Hence, 240 Joule ~ 0.06 Kcal.

d. This amount of mechanical energy has to be subtracted from the calculated 1.2 Kcal released by fermenting 1Brix (see above at **a**) which is (~5%).

To sum up the basic calculation, we can set up as a practical value for the released heat:

(6) | Fermenting **1Brix** will release **1.14 Kcal** as heat

Two examples will clarify the usage of this data.

(1) Assume 20,000L (~5000 Gal) of white must at $22B^o$ and 18^oC (64.5^oF). The wine needs to be cooled and fermented at 10^oC (50^oF). The cooling capacity needed is divided into two sections: (a) to bring down the must temperature by 8^oC, and (b), to maintain it at 10^oC by adsorption of the fermentation heat.

(a) 20,000 x 8 = 160,000 Kcal has to be taken out, say in 10 hours (36,000 sec). The power needed is 160,000/36,000=4.4Kcal/sec. Using the mechanical conversion factor **1KW = 0.24 Kcal/sec** leads us to use a cooling unit of 4.4/0.24 = **18.3 KW**. Or in HP (**1KW = 1.34 HP**), the cooling machinery has to be able to produce for this tank 18.3x1.34 = **24.5 HP**. When a heat-exchanger is used as a very fast method to bring down the white must temperature at a practical rate of say two hours (5 times faster), the cooling power needed is 122 HP (without losses due to bad insulation).

(b) By fermenting 22B° must, each liter will produce 22 x 1.14 Kcal = 25 Kcal of heat (see Brix/Kcal ratio above). Assuming that the specific heat of must and wine are not different than water, and without outside heat transfer, the must temperature will rise up over its initial temperature by 25°C. The whole volume of must will liberate 20,000 x 25 = 500,000 Kcal, which have to be removed in order to keep the must at the same temperature. At low temperature (such as 10°C) the fermentation is slow and quite steady for a long time (average fermentation rate at phase-II is about 1.5B° per day). Assuming a practical length of 15 days (excluding lag-period and declining phase), most of the heat has to be removed at an average rate of 500,000/15/24/3600 = 0.38 Kcal/sec. The cooling unit has to supply power of 0.38/0.24 = 1.6 KW or 2.1HP.

(2) 10,000L of red must with 23B° and fermenting at room temperature (25°C). No pre-fermentation cooling is needed, and the desired fermentation temperature is the same as the initial 25°C. At this temperature the fermentation rate is quite fast and the exponential growth phase will take approximately 5 days, which forces us to calculate the cooling power according to this maximum rate of fermentation, namely, five days. The total heat liberated by each liter of must is 23 x 1.14 = 26.2 Kcal. The whole volume of 10,000L will liberate 262,000 Kcal. In order to control the temperature for 5 days of maximum fermentation rate, a cooling capacity of 262,000/5/24/3600 = 0.60 Kcal/sec is needed, which is 1/0.24 = 2.5 KW or **3.4 HP**. Because red wine fermentation is faster, more power is needed than in white wine to keep the fermentation at the controlled temperature.

A few remarks in regard to the cooling calculations:

(a) The calculations were done for maximum demands at certain stages during the fermentation. On the other hand we completely ignored environmental losses which must be considered specifically in each case.

(b) Another cooling term which is used in the industry is "refrigeration ton," called also ton R. Its conversion to more physical terms can be made by:
1ton R = 3517 KW.

(c) Because the chilling rate is very fast and needs high cooling power, while the fermenting cooling rate is much slower, two separate cooling systems are preferred.

(d) The total cooling capacity of the winery has to be calculated according to the total grape reception rate and fermentation schedule of white and red wines, and the peak demand for quick chilling the white must before fermentation.

Red Wine Temperature

A few words on red wine fermentation. It was realized above how much heat is released during fermentation. In red wine where fermentation is allowed to proceed at higher temperatures, the rate may become so fast that the heat formed in a short time elevates the must temperature to such a level that there is danger to the yeast viability. This is more likely to happen in large volumes where heat transfer is more restricted. It is also important to know that the floating cap has higher temperatures than the must beneath it.

The following figure shows the temperature gradients during fermentation between must and cap:

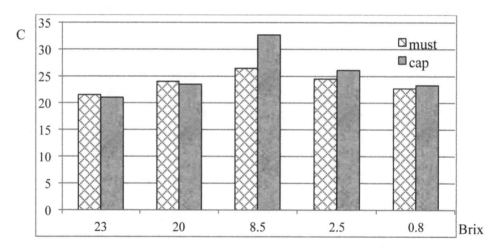

Punching down is important not only to enhance extraction of tannin, but also to equalize the temperatures of the must and the cap.

D. Malo-Lactic Fermentation

1. General Aspects

Malo-lactic fermentation is caused by malolactic-bacteria which belong to three genera: *Lactobacillus, Pediococcus*, and *Leuconostoc*. All these genera may be found in wine, but the species of interest is *Oenococcus-oeni* (former name *Leuconostoc-oenos*), which is the commercial species with strains such as *ML-34, PSU-1, E-145*, and others. These are the "good guys" of the trade, relative to the "bad guys" who belong to the other two genera *Lactobacillus and Pediococcus*.

Chemistry of ML-Fermentation

Malolactic bacteria live on glucose and other sugars and on organic acids, but they also carry on the transformation of malic acid to lactic acid:

$$HOOC - CH_2 - CH(OH) - COOH \quad ---> \quad CH_3 - CH(OH) - COOH + CO_2$$

through three possible pathways:

(1) $HOOC - CH_2- CH(OH) - COOH \longrightarrow HOOC - CH_2- CO - COOH \longrightarrow$
  malic acid $\qquad\qquad\qquad\qquad\qquad$ oxaloacetic acid

$CH_3- CO - COOH + CO_2 \quad ---> \quad CH_3- CH(OH) - COOH$
  pyruvic acid $\qquad\qquad\qquad\qquad$ lactic acid

(2) $HOOC - CH_2- CH(OH) - COOH \longrightarrow CH_3- CO - COOH + CO_2 \longrightarrow$
  malic acid $\qquad\qquad\qquad\qquad$ pyruvic acid

$CH_3- CH(OH) - COOH$
  lactic acid

(3) $HOOC - CH_2- CH(OH) - COOH \longrightarrow CH_3- CH(OH) - COOH + CO_2$
  malic acid $\qquad\qquad\qquad\qquad\qquad$ lactic acid

In this transformation the 4-carbon di-carboxylic acid is decarboxylated into 3-carbon mono-carboxylic acid and CO_2. The term "fermentation" used in this microbial activity refers to the CO_2 formation with no real justification, because no real energy is produced in this process. Thermodynamic calculations show that the free energy difference (ΔG) between products and substrates in the above equation is practically negligible (-2 Kcal/mole), which brings up the question: What kind of benefit do the bacteria gain by malo-lactic transformation? Later study suggests some slight excess of the oxidized NAD^+ coenzyme, which is an important factor in oxidation-reduction metabolism processes that may enhance the bacteria's initial growth rate in the presence of malic acid. Another option is that some extra ATP is formed and is expressed as an increase in lactic acid formation from glucose by malo-lactic bacteria in the presence of malic acid. On the other hand, it may be that the only benefit for the bacteria is the reduction in acidity and increased pH caused by the malic to lactic transformation. The convincing answer is still a mystery.

The reduction in titratable acidity by malo-lactic fermentation (MLF) is in the range of (1 – 3) gr/L, which may be quite important in softening high acid wines. MLF also adds some metabolic products that may improve wine quality. Diacetyl {CH_3-CO-CO-CH_3} is one such major by-product. A survey on Australian and Portuguese red wines revealed an average content of about 3 mg/L of diacetyl in wines which had undergone MLF, compared to an average of about 1.5 mg/L in those which had not.

Benefits of MLF

What are the benefits of having malo-lactic fermentation occur in wine production?

Reduction of acidity – when it is too high, especially in cool regions. Malo-lactic fermentation can reduce the titratable acidity by (1 – 3) gr/L and increase the pH by (0.1 – 0.2) units.

Flavor changes – are expected to result from ML fermentation, as some composition changes occur during this process. Beside the main product, lactic acid, other compounds such as diacetyl, acetoin, 2,3-butandiol, and volatile esters are also produced.

Studies on the question regarding the contribution of MLF to wine quality brought about controversial results. In many cases there was no significant difference between wines with and without MLF. The reduction in acidity is well recognized, and the "buttery" flavor of the extra diacetyl presence in wines that had undergone MLF was also noted. But the contribution to the overall quality and the recognizable difference between MLF wines and those without MLF remains for additional studies.

Microbial Stability – in regard to malo-lactic bacterial infection, stability is gained by having MLF occur during wine processing and not in the bottle. This advantage is significant when the wine conditions are favorable for MLF to occur, namely, high pH, low SO_2 and also in coarse-filtered wines (appropriate for red wines). Aside from these benefits, MLF is not desirable in fruity, fresh style wines (white or red), as the MLF might change the wine's fruity character. Also it is not desirable in low acid wines which after MLF will be more flat and unbalanced.

2. Factors Affecting Malo-Lactic Fermentation

Several factors affect the malo-lactic bacteria growth:

Temperature: Below 15°C (59°F) the growth rate is practically zero. Warmer conditions are needed to stimulate growth. The range of (20 – 25) °C, (68 – 77) °F is optimal.

pH: Malo-lactic bacteria are very sensitive to low pH. At pH = (3.0 – 3.3) *Oenococcus-oeni* can live, but would do better at higher pH. For other genera (*Lactobacillus, Pediococcus*) it is almost impossible to grow at pH below 3.2. At pH above 3.6 they grow well, spoiling the wine by producing a large quantity of diacetyl, acetic acid, and mousy taint. As a result of MLF the TA will be reduced and the pH will increase by an average reduction of 0.15% TA and a rise of about (0.1 – 0.2) in pH units.

SO_2 : Sulfur-dioxide is a very effective inhibition agent against malo-lactic growth, even in its bound form. At pH = 3.5 the presence of 30 ppm free SO_2 was found to completely inhibit malo-lactic bacteria growth and activity. The same concentration of acetaldehyde-bisulfite (bound SO_2) was also very effective, although its action was delayed. The inhibition is effective even at 10 ppm of free or bound SO_2. In other studies the inhibition appears less effective at low concentrations, with concentrations above 50 ppm SO_2 needed to be effective.

Nutrient: Nutrient deficiency can inhibit ML growth. Early racking after yeast fermentation (if MLF was not carried on simultaneously with yeast fermentation) may leave the wine low in vital nutrients, which can prevent the bacteria from growing.

Inhibitors: (a) *Fumaric-acid* {COOH – CH = CH – COOH} is used as an inhibitor for ML bacterial growth when it is not desired. The mechanism of action is not yet clear, but it is very effective when used as an additive in a concentration of (600 – 700) mg/L (at pH = 3.4 where these experiments were done). It is evident that pH and fumaric-acid together have a combined effect on inhibition of ML fermentation. In another study the use of 500 to 1500 ppm of fumaric-acid at three pH values showed the combined inhibition effect of these two factors. Use of 500 ppm, which looks quite effective at pH = 3.5, is certainly not sufficient at pH = 4.1. These measurements were done in wines kept in favorable conditions for MLF (except fumaric-acid), namely, without SO_2, at 25 °C, and without removal of yeast lees after fermentation. In real wine conditions, inhibition would be more effective. So, in clear low pH white wine (3.0 – 3.4) with (20 – 30) ppm of free SO_2, the (500 – 700) ppm of fumaric-acid would be quite effective. If the pH is higher, higher concentrations of fumaric-acid are needed. Addition should not be done prior to yeast fermentation, because fumaric-acid will be degraded by the yeast. So, if MLF is not desired and the use of fumaric-acid is chosen to do the job, it is recommended to add it to the young wine after first racking. Fumaric-acid is difficult to dissolve in wine, which may cause some practical difficulties. Its solubility in water at temperatures of 25°C, 40°C and 100°C, is 6, 10 and 98 gr/L, respectively. The high temperature solubility is recommended for use when needed. Fumaric-acid showed no negative effect on wine sensory score when added up to about 1.5 gr/L.

(b) *Fatty acids* are known to be microbial toxic factors. Their inhibition activity on malo-lactic fermentation in wine range concentrations showed to be definite effective, especially a mixture of hexanoic, octanoic, and decanoic acids. As individual acids, their inhibition was much less effective. Dodecanoic acid (C_{12}) was also found to be an effective ML inhibitor.

3. Inoculation of Malo-Lactic Fermentation

Induction of MLF can be spontaneous or by inoculation. Spontaneous MLF canhappen at any time when the right conditions are met (from fermentation through all vinification processes and even in the bottle).

The easiest and most convenient way to initiate MLF was developed in the last two decades by using direct inoculation of dried bacteria into the wine. The ML commercial material is sold as lyophilized freeze-dried bacteria ready for use right after re-hydration. The selected *oenococcus-oeni* strain can survive the direct introduction of the bacteria into the wine conditions at any time from the yeast's full activity during fermentation until any time afterword. When using the commercial freeze-dried bacteria, the following notes can be helpful:

(1) The dried bacteria should be kept in low temperature (in the freezer) prior to its usage. At such conditions its shelf life is about one year.

(2) The ML powder may be applied either directly to the wine tank, or be re-hydrated with distilled water (20 – 30) °C at volume 30 times of the bacteria weight for 15 minutes and then introduced it to the wine with minimal aeration.

(3) The total SO_2 concentration should be below 25 ppm and less than 10 ppm of free sulfur-dioxide. This factor is of major importance.

(4) The wine temperature should be higher than 20°C.

(5) The recommended quantity of bacteria dry powder is 25 gram/2500L.

(6) In certain cases where a nutrient is known to be deficient, or even to ensure a fast MLF, it is recommended to add special ML nutrient supplements which are

sold by the ML bacteria producers.

(7) Also each commercial producer of the ML bacteria has his own specific instructions which might be different in some aspects from those given here. It is advisable to read them.

It is very important to finish the MLF as soon as possible! The reason is that SO_2 cannot be added until it is done, leaving the wine unprotected against other microorganisms, mainly *Brettanomyces* yeast, *Acetobacter*, and the "bad guys" of ML bacteria (*Lactobacillus and Pediococcus*). So the MLF should be finished soon, and the SO_2 should be added to the fresh young wine.

As for the timing, in cases where it is not left to occur spontaneously, MLF can be initiated simultaneously with the yeast fermentation, or after it is over. Post-fermentation MLF can be more difficult to maintain for several reasons: high alcohol concentration, low nutrient content, some SO_2 added after fermentation or produced by the yeast. Besides, there is some advantage to finishing both fermentations at the same time. Therefore, simultaneous yeast fermentation and MLF has become very common in recent years. The bacteria have access to fresh must nutrients and low alcohol at the start. This enables the wine to finish the MLF more or less at the time of alcoholic fermentation with less risk of other microbial spoilage. Then SO_2 can be added right after the two fermentations are over. When MLF is carried out simultaneously, SO_2 should be avoided during fermentation, or be added in less than 25 ppm.

After presenting the whole advantages of simultaneous fermentations, it should be noted that there are some questions on this matter. First, there is a possibility of having higher volatile acidity in the case of simultaneous fermentation, especially when yeast fermentation is somehow delayed. This is because of sugar consumption by the ML bacteria, which produces acetic acid as by-product. And if the fermentation is stuck, the result might be really a serious problem. It is much safer to wait until the fermentation is over and the wine is dry (less than $(2-3)$ g/L of residual sugar) before inoculating with ML bacteria. On the other hand, you have to leave the fresh wine unprotected by SO_2 until the MLF is finished, which is also quite risky. So, there is no clear answer whether to do simultaneous fermentation or to do it in tandem, namely, first alcoholic and then malo-lactic.

In this connection, it is worth mentioning in more detail the effect of SO_2 production during alcoholic fermentation by the yeasts. Practically all yeasts produce SO_2 in a measured concentration range of $(10-40)$ ppm total. At this SO_2 range the ML bacteria are not strongly inhibited. The situation is more severe when SO_2 is introduced initially into the must before fermentation, so it accumulates with the SO_2 produced by the yeasts. The result of such mutual fermentation of yeasts and ML bacteria is that there is no practical advance of MLF until a very long period after the alcoholic fermentation has ended. Various studies showed different results to the question of whether simultaneous fermentation may finish the MLF earlier than sequential alcoholic-ML fermentations. It was also shown that the yeast strain used for the alcoholic fermentation strongly affects the so-called "malo-lactic fermentability" after finishing its job; in other words, leaving the new wine in such condition as might reduce the following MLF rate. The difference lies in the yeast's SO_2 production and the pH after alcoholic fermentation.

To sum up this issue, although simultaneous fermentation is known and practiced in the industry, it seems to be better to do it one after the other.

When MLF is *not desired*, measures have to be taken to prevent it. Otherwise, it will probably happen. In the winery, the best measure is low temperatures (below $10^{\circ}C$) and a sufficiently SO_2 level, above 25 ppm free. MLF in the bottle is a very serious problem. It will spoil the wine by CO_2 bubbles, sediments, and probably a bad smell. Before bottling, enough SO_2 should be added and sterile filtering at 0.45μ should be employed to assure that the malo-lactic bacteria will not start fermentation in the bottle.

Following Up ML Fermentation:

There is a very simple analytical method to determine whether MLF has occurred and if it is finished. This test is a qualitative determination only, but for all practical purposes it is good enough to serve its purpose.

The test is performed by paper chromatography with a special eluting solvent, which is made from the following mixture: n-butanol (100 ml), water (100 ml), formic acid (10.7 ml), and indicator solution (15 ml). The indicator solution is 1% of *bromocresol-green* in ethanol. The mixture's components are shaken and placed in a separator funnel. The aqueous phase (lower one) is discarded from the funnel bulb. The aqueous phase has to be separated from the organic phase by several discards, until no water is seen at the bottom of the funnel.

A drop of the tested wine is spotted on the special chromatographic paper by a capillary tube. The eluting mixture is placed in the bottom of a close jar to about (1 – 2) cm height, and the paper is placed in the eluting solvent, wetting its bottom.

The jar can be covered. After a while (about two hours) all the spotted wine acids will be moving upwards on the paper by the eluting solution. When the climbing liquid level reaches the top of the paper (some few cm below it), the paper can be removed from the container and allowed to dry. After a few hours, a chromatogram is developed on the paper, showing the location on the paper of the major acids present in the wine. To enhance the color contrast of the paper chromatogram , move the dry paper above an open bottle of ammonia for a second. The paper will turn blue, leaving white spots in the places of the wine acids.

A reference chromatogram of the major wine acids (tartaric, malic, lactic) can be prepared and run on the same paper with the wine samples, to indicate each acid's position on the chromatogram. The chromatogram will indicate whether malic acid is still present in the wine or not.

CHAPTER IV

CELLAR OPERATIONS

Chapter iv: Cellar Operations

A. Racking

Wine is racked off a couple of times during processing. Racking is another word for decanting, meaning transfer of liquid above its solid sediments. In wine processing, solids are called *lees* and may contain yeast cells, pulp, tartrate crystals, fining particles, proteins, and tannins. In the racking operation, two major changes occur: *clarification* and *aeration*.

In general, the main operational principle in racking is simple: *as little as possible*. This means that the racking should be done only when certain operations are necessary, such as leaving sediments at the bottom of the tank or when blending the wines, or any wine transfer that is needed during processing. Wine tanks contain a few outlets with valves. The bottom one can rack off all the liquid from the tank. The second valve is located about (20 – 30) cm above the bottom one. In most cases the lees layer is somewhere below the level of the second valve. Therefore, racking off the wine (or white juice) from its solids content is done from this valve. When air starts to be seen at the hose, the racking is over and the pump should be shut off. The rest of the wine may contain a low or a very high quantity of solids. In the latter case, either the bottom is drained off, or some of the wine can be saved. The best way to save this wine is to let it out from the bottom valve to a small container and let it rest for a day before racking it again above its solid layer. When the sediment layer is very low and dense, in order to save extra work, the hose can be connected to the bottom valve and the rest of the wine carefully racked off. Some portion of the sediments will be transferred into the racked wine. Some tanks have the option of an inner bent tube placed in the second valve, inside the tank, which can be rotated from the outside by the operator to alter the level of suction. By gradually lowering the inner tube, it is possible to pump out the wine exactly to the lees surface level. In this way, the racking is continued until this point, and the rest is drained off. The lees layer can be watched from the tank's top opening. Two people are needed for this operation. This device is very helpful and is highly recommended. The other side of the hose is connected to the receiving tank, either at the bottom valve if no aeration is desired, or at the top of the tank when aeration is desired. All connections between tanks, hoses and the pump should be well connected, with no leakage of liquid out, or leaking in of air.

The racking operation can also be used for introducing and mixing some necessary materials that are used during wine processing, such as fining agents, sulfur dioxide, acids addition, or blending with another wine. When this is the case, in order to complete the mixing, the wine can be circulated in the tank from bottom to top. Any aeration during racking introduces oxygen into the wine at concentrations from tenth of mg/L in careful racking, up to (5 – 6) mg/L of oxygen in over-aerating racking. For white wine, less oxygen is better, to prevent browning and preserve freshness. In red wine, certain amounts of oxygen are necessary during the process of aging, so racking with some aeration is a good practice. In order to minimize oxygen absorption during racking, the receiving tank can be filled prior to racking with CO_2 gas through the bottom valve for a couple of minutes. The CO_2, which is heavier than air, will stay on the wine surface and will protect the incoming wine when it fills the tank. Nitrogen is another good option. It is completely unnecessary to use argon for this, as some wineries do. It is very expensive, and for the purpose of preventing oxygen from being dissolved in the wine, nitrogen or CO_2 will do the job. The major rackings are:

Zero Racking: In white wines. We call it zero racking because it is done before the fermentation. This is the clear juice racking from the must lees which is necessary for good white wine making.

First Racking: After fermentation has terminated, this racking roughly clarifies the new wine from the very dense sediments which contain mostly yeast cells and pulp. It should be done shortly after fermentation has stopped, to prevent extraction of dead yeast constituents into the wine, like protein metabolites, amino acids, and hydrogen sulfide. In red wine, there is a large quantity of seeds in the lees, and if left un-racked for a long time, the wine extracts much of the tannins. In certain styles of wines, both whites and reds, the first racking is delayed for a long time, regardless of the above considerations. In white wines it is done especially in selected lots of grapes (Chardonnay or Sauvignon Blanc) which are barrel fermented and left after the fermentation for a couple of months on the lees (*sur lies*). In this case the yeast is mixed every week or so with the wine, and its effect on wine character and quality is very noticeable. The main effect is that it lessens the oaky nose of wine that stays in a new barrel for a long time, especially white wine.

Second Racking: In white and red wines, the second racking is needed to further clear the wines from the sediments left over from the first racking.

Third Racking: In white wines, after fining (with bentonite) and cold-stabilizing, which can be done consecutively and the wine then racked-off at once. In this case the lees consist mostly of fining particles and tartrate crystals. This racking can also be a good chance for blending. In reds, the wine is racked to barrels, to start its barrel aging.

Fourth Racking: In whites, this racking is for the first filtering. Because filtering is in fact transferring wine from one tank to another one through a filter, we consider filtering as racking. In certain cases filtering can be done right after bentonite and cold stabilizing, namely, in the third racking. In this case, the fourth racking can be avoided. The racked wine is then stored in tanks until bottling. In reds, during barrel aging, racking from barrels to other barrels is done at intervals of four to six months. In this racking the barrel's bottom sediments, which contain yeast cells, fine pulp particles, bi-tartrate crystals, and some polyphenols-protein colloids, are cleared off. The empty barrels are washed with water and refilled from other racked barrels. In many cases it is done differently, namely, by racking all the barrels into a tank, cleaning the barrels, and then racking the wine back into the barrels.

Fifth Racking: In white wines, from the tanks to the bottling line through a sterile filter. In reds, from the barrels back to the tanks to unify the barrel's wine and to blend it for further storing and aging before bottling.

Sixth Racking: Red wines are racked off from the tanks to the bottling line through a filter.

The above racking scheme represents a general framework. Because the wine is exposed to oxygen in each racking, addition of SO_2 during each racking is advised. The actual quantity can be determined by measuring the free sulfur-dioxide before racking, then adding enough to have in the racked wine about $(20-25)$ ppm of free SO_2 after racking. (For details look in chapter VII, section C.) The small additions of SO_2 at each racking will accumulate to be the total SO_2 in the wine.

B. Stabilization

In the wider sense, this term refers to operations that prevent cloudiness and settling of particles in the bottle. The causes of cloudiness and solid particles in the bottle can be protein coagulation, polyphenol colloids, protein-metal haze (iron and copper), and tartrate crystallization. Protein and polyphenol stabilization will be discussed in the "Fining" section. Iron and copper haze, are very rare now, as all cellar equipment is made of stainless steel; they will be discussed in details in chapter VII, section D. Here we shall deal with tartrate stabilization.

1. Tartaric-acid dissociation

Tartaric acid (represented here by H_2T) is the major and unique acid in grapes and wines. It dissociates into bitartrate HT^- and tartrate $T^=$ ions and forms salts with the major cation in wine, potassium, to form potassium-bitartrate salt **KHT**:

$$H_2T \overset{K_{d1}}{\Longleftrightarrow} HT^- + H^+$$

$$HT^- \overset{K_{d2}}{\Longleftrightarrow} T^= + H^+$$

Three components are at equilibrium, namely, undissociated acid (H_2T), bitartrate-ion (**HT** $^-$), and tartrate-ion (**T** $^=$). Their concentrations at equilibrium are connected by the equilibrium constants K_d and their logarithm (with minus sign) pK_a. In the case of double dissociation (tartaric acid), there are two sets marked as 1 and 2. The K_d values and their corresponding pK_a (at 25°C) are:

$$K_{d1} = 9.1 \times 10^{-4} \quad ; \quad pK_{a1} = 3.04$$
$$K_{d2} = 4.25 \times 10^{-5} \quad ; \quad pK_{a2} = 4.34$$

Based on these pK_a's and using the formulation for weak acids (see Pre-Harvest chapter, section A), the distribution of tartaric acid components (HT, HT $^-$, T $^=$) in water solution, at wine pH range can be calculated, as is shown in the following figure (note the maximum concentration of bitatrate ion at pH = 3.7, which will be discussed later):

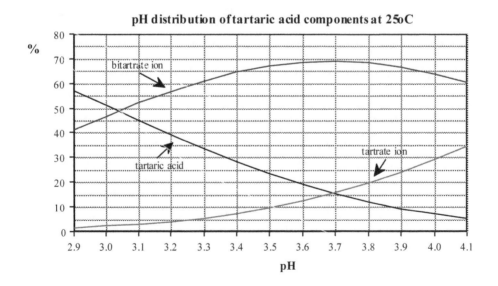

pH distribution of tartaric acid components at 25oC

2. Potassium-bitartrate

The potassium-bitartrate salt COOH-CH(OH)-C(HOH) - COOK, also called *cream-de-tartar*, forms crystals which belong to the orthorhombic space group. The crystals are soluble in grape juice, but are less soluble in the alcoholic solution formed after fermentation.

Any ionic salt will precipitate when the *concentration-product* (CP) of that salt exceeds its *solubility-product* (SP) in solution at the specific conditions under study. The *concentration-product* is the product of the negative and positive ions' molar concentrations in solution. The *solubility-product* is the product of the ions' concentrations in solution, when the salt is in its maximum possible concentration, namely, when it is saturated. In formal presentation, a salt in solution dissociates to its ions:

AB <===> A$^+$ + B$^-$

and   **CP = [A$^+$] x [B$^-$]   ;   SP = [A$^+$]$_{sat}$ x [B$^-$]$_{sat}$**

where [A$^+$] and [B$^-$] are the A$^+$ and B$^-$ molar concentrations of the ions.

SP is equal to CP when the salt is "saturate" in that solution.

There are three possible relations between CP and SP:

(1) When **CP < SP** the salt concentration is below its saturation point. Then, the salt in solution is stable and will remain so, as long as the conditions are unchanged.

(2) When **CP = SP** the salt in solution is at its saturation point, a state where the solution can not contain any more of the solute in solution, unless the conditions are changed to make it possible.

(3) When **CP > SP** the salt in solution is over-saturated, an unstable solution (called super-saturation), in which any minor change in the solution conditions will cause the solute to crystalize and precipitate.

Let us check the case of potassium-bitartrate in wine. The concentration-product of potassium-bitartrate in wine is defined as:

CP = [K$^+$] x [HT$^-$] = [K$^+$] x [total H$_2$T] x % [HT$^-$]

The wine pH, its alcohol concentration, and the temperature are the most important factors affecting the solubility of potassium-bitartrate in wine:

pH: We showed in the figure above the various distributions of tartaric acid components vs. pH values in a given wine pH range. As can be seen in the figure, as wine pH gets higher, the tartaric acid portion gets lower, and the percent of bitartrate ions gets higher, which makes the concentration-product of potassium-bitartrate become larger too. For example, at pH = 3.1 the bitartrate ion is 52% of the total tartaric acid present, while at pH = 3.4 it is 65%. Hence, at higher pH, the concentration-product of tartrate is higher too. This means that at the same *molar* concentration of potassium bitartrate in two wines, the one with the lower pH may be stable, while the one at higher pH might be unstable and precipitate (according to principle (1) above).

Alcohol: Potassium bitartrate is quite stable in grape juice, although its CP is close to its SP (below it, of course). During wine fermentation, it becomes unstable and starts to precipitate in the presence of alcohol in the wine. The reason for the precipitation is an effect known as salting-out. In this effect, the less-dielectric solvent (alcohol) added to water reduces the solubility-product of ionic substances, causing them to precipitate if their concentration-product is close to their solubility-product. In this case the SP of potassium bitartrate became smaller in the new medium, wine. And from the previous state where CP < SP, a new state where CP > SP happens, causing part of the salt to precipitate, until equilibrium CP = SP is established.

Temperature: Also influences strongly the solubility-product of any salt, including potassium bitartrate. The solubility-product at high temperatures is higher than at lower temperatures. This means that the salt can change its solubility status from CP < SP to CP > SP just by reducing the temperature of the solution.

All these factors, namely, the pH, the alcohol content, and the temperature, affect the solubility and therefore the wine stability towards tartrate precipitation. In red wine, other factors also influence the solubility of potassium bitartrate, such as other organic acids and polyphenols in red wines, that inhibit the precipitation. (This is why in old red wines which were stabilized in the winery before bottling, there are always crystals of potassium bitartrate. As the phenolic compounds polymerize and therefore their content is reduced after long aging, their inhibition of potassium bitartrate crystallization diminishes, and fine crystals of the salt are seen on the cork and in the bottle.)

3. Cold Stabilization

The concentration ranges of tartaric acid and of potassium ions in ripened grapes are about (2 – 6) gr/L and (1 – 2) gr/L respectively. The concentration of potassium bitartrate in grape juice is close to its saturation, and as the fermentation proceeds (forming more and more alcohol), the salt becomes *supersaturated*. In such a case, precipitation is inevitable.

Deposition of potassium bitartrate in the bottle is considered an aesthetic defect, and measures are taken during vinification process to eliminate this instability potential. The most common method for tartrate stabilization is chilling the wine to such a low temperature, that the concentration-product of the salt left over after precipitation will be much less than the solubility-product at normal environment temperatures. Therefore the wine will reach stability. When a cold stabilization takes place, changes in wine acidity and pH occur as well. These changes are caused by the change in the concentration of tartaric acid and potassium ions. The acidity is lowered because tartaric acid is removed from solution. But the pH changes depend on the initial pH, where the key factor is **pH = 3.7**. The above figure of tartaric acid dissociation vs. the pH shows that at this pH the bitartrate concentration is at its maximum. When the initial pH values is *lower* than 3.7, any reduction in the bitartrate concentration brings back the equilibrium curve of bitartrate by *lowering* the pH. For example, if the initial pH was 3.4 and the reduction in bitartrate concentration by chilling was from 65% to 60%, then the bitartrate equilibrium line at 60% is at pH = 3.3, which is lower than the initial one by 0.1 pH unit.

On the other hand, if the initial pH is *higher* than 3.7, any reduction in the bitartrate concentration brings back to the equilibrium curve by *increasing* the pH. For example, if the initial pH was 3.8 and the reduction in bitartrate concentration by chilling was from 69% to 65%, then the bitartrate equilibrium line at 65% is at higher pH = 3.95.

As a result, when stabilization occurs, the pH changes according to the initial pH value, namely, below pH = 3.7 the pH will decrease, and above it, the pH will increase. But, in the pH range closely to 3.7, where the curve is almost flat from, say, pH = 3.65 to pH = 3.75, there will be practically no change in the pH after chilling.

There are some formulas and tests that can be done to determine the optimal temperature and duration of wine chilling. The time needed for achieving stability depends on the kinetics of bitartrate crystallization under cold stabilization conditions, namely, the temperature and the initial concentration-product of potassium bitartrate, and on nucleation factors which may increase or decrease the crystallization rate. The initial concentration-product depends on the potassium-ion concentration, the tartaric acid concentration, the pH, and the alcohol content. In general, crystal growth rate in solutions is controlled by two parameters: the transportation of the ions from the solution to the growing crystal (diffusion), and the rate the solute ions migrate on the crystal surface to their final position in the crystal. Diffusion rate is mainly determined by the temperature, while the surface migration depends on the crystal particle's size and impurities which may enhance or block the crystallization process. These "impurities" are wine colloids such as proteins, fining agents, polyphenols, pectin, and other hydrocarbons. They may interact with the K^+ and the HT^- ions, blocking their way on the crystal surface.

On the other hand, the impurity particles may serve as nucleation sites for the supersaturated solution to crystallize. It is therefore not accurate to predict the stability of wine just from its concentration-product, as could be done if the medium were just an aqueous-ethanol solution. We think that this is of no practical importance, because simply by keeping the wine at a temperature range below 0°C and above the freezing point (-5°C), for two to three weeks, the potassium bitartrate stabilization will be set. The cooling can be done either by passing the wine through a heat exchanger (chiller), resulting in quick cooling to the desired range, or by lowering the temperature of the tank by its temperature control, which will take a couple of days. In our opinion, the latter option is preferred, especially in white wine, because it eliminates the need for one more wine transfer. To ease the crystallization and precipitation, it is advisable to combine bentonite treatment (to be discussed later) with the stabilization operation. In such a combination, the bentonite is added with the right amount of sulfur dioxide (25 – 35 ppm), mixed well, and then after a day, the temperature control can be set down. The fining particles will facilitate the crystals settling. Cold stabilization can be hastened by seeding with crystals of potassium bitartrate. Most of the potassium bitartrate precipitates on the side walls and the bottom of the tank.

After one to two weeks, the wine can be racked off and cold filtered before bottling. The winemaker saves one racking, since these two operations (fining and stabilization) would otherwise be done separately. It is possible to check the TA and pH before and during cooling. Following the changes can give some clue when settling has come to equilibrium. Also, the bitartrate crystallization rate can be monitored by measuring the electrical conductivity of the wine. It will continually reduce as long as the bitartrate concentration is reduced, making it possible to determine the end of the crystallization.

4. Calcium Tartrate

Calcium tartrate (CaT) is less soluble in wine than potassium bitartrate. And because the calcium concentration in wines ranges between (30 – 200) mg/L, it may contribute to tartrate instability. The sources of calcium in wines are from its natural content in grape juice, and from additions to the wine by various agents containing calcium, such as $CaCO_3$ (for acid correction), bentonite, filter-aid, or by storage in concrete tanks. When the concentration-product of calcium tartrate in the wine is greater than its solubility-product, it will precipitate, although its precipitation rate is very slow and may take months to achieve stability. Also, racemic calcium tartrate (DL-tartaric acid) is much less soluble (about 35 mg/L in water at 20°C) than the L-calcium tartrate (about 300 mg/L at 20°C). The natural grape juice L-tartaric acid slowly undergoes some racemization over time, and therefore, calcium tartrate, which is considered stable, might become unstable after certain time. This is because its concentration-product exceeds its new solubility-product, which is about eight times lower. This fact can be used in stabilizing wine with regard to calcium tartrate. Addition of racemic DL-tartaric acid will cause faster precipitation. The practical addition dose is about three times the calcium concentration (at the range of 100 to 500 mg/L of DL-tartaric acid).

In white wine, if consumed young, within 1 to 2 years, there should not be a problem. In red wine the tartrate is more stable in solution because of the tannins. And in any case, some sediment in red wines after many years in the bottle is tolerable and accepted.

5. Inhibition and Promotion of Tartrate Crystallization

The function of *inhibition* agents in this case is to prevent the treated wine from undergoing bitartrate precipitation, thereby making it stable in this respect. The principle of their action is to interact with the crystal surface in a way that prevents any further growth. By doing so, although the wine is supersaturated, it will remain stable. Cold tests done on treated wines with inhibition agents do not show any tartrate precipitation, and therefore they are considered stable.

In contrast to inhibition agents, *promoting* agents aim to increase the crystallization rate of bitartrate at room temperature and thereby increase the wine's stability. The main advantages of using such agents (inhibitors and promoters) are that stabilization can be achieved immediately with no cold stabilization (saving energy costs), and it can also assure stability of the wine for a certain period of time.

Metatartaric acid: Addition of very small quantities of metatartaric-acid (about 100 mg/L) inhibits crystal formation in wines for a certain time. The inhibition is a result of interaction of this substance with the growing crystal, preventing any bitartrate ions from joining the surface. Metatartaric-acid is made from tartaric acid by heating it above its melting point at 170°C to form a polymer of tartaric acid. After cooling, the solid product is ground to form a fine powder. Before adding it to the wine (prior to bottling), it should be dissolved in cold water (about 200 gr/L, and added at 100 mg/L to the wine).

After addition to wine, the metatartaric-acid is slowly hydrolyzed back to tartaric acid, losing its inhibition activity. The efficiency and duration of inhibition depends on the storage temperature. The cooler the wine, the longer precipitation can be inhibited. At room temperature it may last up to one year. The use of this agent to stabilize wine from bitartrate precipitation is therefore good for short-aging wines such as whites, rosés, and light reds.

Carboxymethylcellulose (CMC): Another colloidal substance that interacts with tartrate crystals and inhibits their growth very efficiently is the polymer *carboxymethylcellulose*.

This substance is a cellulose chain, in which some of its hydroxyls (OH) are partially substituted by sodium-carboxymethyl groups. The length of the polymeric chain (25 – 250) units, and the number of sodium-carboxymethyl groups per sugar monomer (up to three such substitutions per each sugar monomer) determine the solubility of the polymer in water. The commercial substance has good solubility in cold and hot water, and forms a very viscous solution at quite low concentrations. It is approved as a food additive and it is used mainly in liqueur production to make it thicker and smoother at concentrations of about (1000 – 2000) mg/L.

In wine it acts like metatartaric acid in stabilizing the wine from bitartrate precipitation, but it has the advantage that it is very stable and does not lose its efficiency with time and higher temperatures. It was found to be effective at very low concentration ranges (25 – 50) mg/L. Wines treated by this agent show very good stability in cold test.

Calcium seeding: This agent contains calcium tartrate and calcium ion in a powder form. When added to wine, the calcium ion reacts with the excess tartrate ion (T $^=$) to form insoluble calcium tartrate crystals. The calcium tartrate particles in the reagent function as seeding nuclei for crystal formation. The concentration range for using this agent (commercially known as "Koldone") is between (0.5 – 2.5) gr/L. The exact quantity can be found in a laboratory test by increasing concentration of the agent in samples of the wine. The agent is mixed well with the wine samples (at room temperature) and left for 24 hours. Then the wine samples are filtered and placed in cold temperature (from -3°C to -8°C) for about 10 hours (freezing may occur). After a day at room temperature the samples are checked for crystal formation. The sample in which the crystals were not formed, with the minimal amount of the agent added, is the one to use for stabilizing. For convenience, the agent is used as a 12% slurry in water. After preparation, the slurry should be left overnight and remixed just before adding it to the wine, while mixing the wine well by pumping. The wine should then be left to settle down, and a stability test should be done. If the wine is stable it can be racked and filtered.

The last thing to remember is that after blending wines, there might be instability potential, and a cold stability test is recommended.

C. Fining

Fining is an operation of adding certain substances into wine, in order to improve its appearance (turbidity, color), to remove some off-flavors, and to prevent some potential instability which may be caused in the future by some wine components. The fining agent reacts with the specific wine component either chemically or physically, to form a new component that tends to separate from the wine either by precipitation or filtration. Because the interaction is not always of a purely chemical nature, the quantity ratio between wine components and the fining agents is not strictly constant, and a preliminary lab test is necessary in each case prior to the addition of the agent. In certain cases the fining agents may themselves contribute either an off-flavor or add instability to the wine, or remove and reduce some of its natural flavor. Fining agents must therefore be used carefully and with the minimum necessary doses, in order to avoid over-fining.

1. Protein Stabilization

The protein content of grape juice varies in the range of (100 – 800) mg/L. During fermentation the protein content may either increase or decrease (up to about 40%). The amino acids in proteins act as amphoteric dipolar ions. They can be represented by the dipolar formulation $NH_3^+ - CH(R) - COO^-$ which means that they behave as protonated ($NH_3^+ - CH(R) - COOH$) in low pH, or as anionic ($NH_2 - CH(R) - COO^-$) in high pH solutions:

$$H^+$$
Acidic solution $NH_3^+ - CH(R) - COOH$ <======> $NH_2 - CH(R) - COOH$

$$OH^-$$
Basic solution $NH_2 - CH(R) - COOH$ <======> $NH_2 - CH(R) - COO^-$

The dipolar ion behavior of protein is a result of the association/dissociation reactions of its amino acids in an acidic/basic medium. At wine pH (acidic) the protein's amino acids are therefore positively charged (upper part of the equations). The solubility of proteins in an aqueous medium depends on the pH of the solution and the *isoelectric point* (PI) of the protein. The *isoelectric point* is the pH at which the positive and the negative forms of the protein are at equal concentrations. At this specific pH the protein is electrically neutral, and therefore it is *least soluble* in aqueous solutions.

Wine pH range (3.0 to 4.0) is very close (slightly below) to the isoelectric points of most of its proteins. Also, the alcohol in wine reduces the proteins' solubility. These two factors may cause some protein instability in wine, and this means that precipitation of the proteins is only a matter of time, mainly at higher temperatures. The result is a colloidal cloudiness and haze. Sauvignon Blanc wines, for example, are known to have high concentrations of unstable proteins. The protein colloidal particles that are positively charged may interact (under certain conditions) with negatively charged particles to form neutral particles that can coagulate (flocculate) and settle down. This is the principle of protein removal from wine by fining. The two major agents used for that purpose are *bentonite* and *silica-gel*.

***Bentonite*:** Bentonite is a mineral material composed of aluminum-silicate anions ($Al_2O_3 \cdot SiO_2$) $(H_2O)_n$ neutralized by cations such as calcium, sodium, potassium, and magnesium. Sodium-bentonite is the most common and effective, but calcium-bentonite is also used. The microscopic structure of the mineral reveals very small plates of the order of 0.1μ with a very high surface area of about

(500 – 1000) square meters per gram of bentonite. The plates can absorb water and swell to form a colloidal suspension. Being negatively charged, bentonite particles can interact with positively charged protein molecules on its plates' surface, and consequently the protein is removed from the liquid by precipitation of the heavy bentonite particles. The interaction of bentonite with proteins is very fast, in a matter of minutes. Afterwards, the adsorbed proteins slowly settle by gravitation of the bentonite particles to form bentonite lees at the bottom of the tank. The mechanism of adsorption between the protein and bentonite is assumed to be of a fixed number of absorption sites on the solid bentonite plates, which bind to the protein molecules on a one-to-one basis. This leads to describing the bentonite-protein interaction by the *Langmuir equation* of adsorption, as was shown by a study in a model wine.

This is the common case of protein removal. But in those cases where the protein segments have isoelectric points which are very close to the actual wine pH (hence they are neutral), their interaction with bentonite is minimal (close to zero), and therefore such proteins are very difficult to remove by bentonite. Unfortunately, these proteins are also the most unstable in solution and are the most likely to cause protein haze. Also, there are protein fractions that have their isoelectric points *below* the actual wine pH, and therefore they are *negatively charged* and consequently will *not* interact with bentonite at all. Fortunately the number of such fractions is very small. Because of these considerations, it is worthwhile to keep in mind as a general guide, that the lower the wine pH, the higher is the efficiency of protein removal by bentonite. Or in other words, it should be understood that the use of bentonite to remove proteins from wine has some limitations. And in cases where high quantities of bentonite are used in order to achieve stability, the wine might be partially stripped of its flavor components, gaining an earthy taste, with an over-fined wine as a result. Fining with bentonite is done for protein removal from *white wines* only. It does not make sense to do so in red wines, as will be discussed in the next paragraph (on phenolic fining).

When the fining is done on white *grape juice* to facilitate clarification, the reduction in proteins and amino acids may cause fermentation problems (lower fermentation rate, stuck fermentation, high residual sugar). It may also affect and increase the formation of hydrogen-sulfide.

Preparation of bentonite for use is done by hydration of its particles in hot water (80 – 90) °C, for at least several hours (preferably one day) before adding it to the wine. The recommended concentration is **5% (weight / volume)** for example

1kg bentonite in 20L of water. The hydrated bentonite swells and forms a very viscous slurry gel.

The amount of bentonite used should be minimal but enough to fulfill its duty. An excess of bentonite reduces color and aroma, and may leave some off-flavor. One should test in the laboratory for the right amount of bentonite to use by testing a series of wine samples (say 0.5L each), with increased quantities of added bentonite: (0.2/ 0.4/ 0.6/ 0.8/ 1.0) gr/liter, equal to (0.4/ 0.8/ 1.2/ 1.6/ 2.0) liter of slurry/HL of wine. All samples (with untreated one as a control) should be well mixed with the bentonite slurry and allowed to stand overnight. After settling, the samples should be decanted, filtered with one micron filter, and checked for protein stability. The test is done as follows: the clear samples are placed in a thermostatic bath at 90°C for two hours, and then at 4°C for six hours. After coming back to room temperature the samples should be checked for turbidity appearance. The unstable proteins in those wine samples with not enough bentonite added will form a noticeable haze. The sample with the lowest bentonite concentration that is clear is the one to use in this wine. In most cases it falls between (0.2 – 1.2) g/L. After treating the whole batch of wine with the selected bentonite concentration, it should be left to settle down in the tank for a week. Then it is recommended to take a sample from the tank, filter it, and test it for assurance of its stability. If haze is seen in the tested wine, it is still not stable, and some more bentonite should be needed. If the test shows that the wine is clear and stable, the treated wine should be racked off from the bentonite lees.

When added to the wine, bentonite acts best at moderate temperature (15°C to 25°C). The sediment volume of bentonite is about (1 – 2) % of the wine volume, which may cause some wine loss. To minimize the loss, it is therefore advisable to treat first with bentonite and then to follow with the cold stabilization treatment. The recommended procedure is as follows: bring the wine temperature to (15 – 25) °C and add bentonite. Allow the sediments to settle down for a few days, and then do the cold stabilization and filter it (cold). The seeds of potassium tartrate which are formed during cold stabilization will help to reduce the bentonite lees level to a minimum.

Some varieties, such as Sauvignon Blanc or Gewurztraminer, need some additional bentonite in order to be protein stabilized. Excess use of bentonite (over 0.8gr/L) may strip some of the aroma. In certain cases, a compromise must be made between prospective protein instability and reduction in the wine's aroma.

Silica-gel: Silica-gel fining agent is an aqueous colloidal suspension of silicon-oxide prepared from sodium silicate ($Na_2SiO_3.H_2O$) by addition of strong acid to form silicic acid. The amorphous silicate particles (typical diameter of 50 Angstrom) are held in suspension by the negative charges on its surface. The charge is the result of the dissociation of the surface hydroxyls, but it is also stabilized by sodium ions added to it:

$$\begin{array}{c}
\overset{\displaystyle O^-}{\underset{\displaystyle /}{|}} \qquad \overset{\displaystyle OH}{|} \\
-\,Si-O-Si \\
\end{array}$$

The commercial silica agent (called *kieselsol, baykisol, klebsol*) is a milky-like suspension of 30% (by weight) of SiO_2. The negatively charged particles can interact with positively charged protein molecules, the same as bentonite does, or as tannin can do. The combined action of the two agents, *silica-gel* and *gelatin* (which will be described later) in clarifying cloudy wines is very efficient and common. Silica is a very clean fining agent in the sense that it can be made at high purity and does not contain other components that might affect the wine aroma and flavor. When it is used for white wine clarification and stabilization, addition levels of silica are in the range of (0.1 – 0.25) ml/L (when using the commercial 30% liquid form). When added with *gelatin* to white wines, the recommended range of the *gelatin* addition is (20 – 50) mg/L. In such case, the two agents should be added to the wine separately, the silica being added first. Coagulation and separation from the wine begins in a very short time after addition. If bentonite is also used, it is better to add it first, and then the silica followed by the gelatin. Excellent clarity can be achieved with silica-gelatin combination in difficult cloudy cases such as botrytis wines. The use of silica for clarification is also common in other beverages. Silica agent was approved for use in foodstuffs in the US by the FDA and by the International Oenological Codex.

The main advantage of using silica for fining is as a replacement for tannin in the gelatin fining of white wine. The tannin is added in order to avoid excess gelatin content in the fined wine, but it may also add some astringency to it. The silica is neutral with no taste effect at all, so over-fining is not a risk.

2. Tannin Reduction

The presence of tannins in red wines at high concentrations may lead to astringency and a hard mouth-feel. In white wines they may sometimes cause bitterness, and after being oxidized, a yellow-brown color. Therefore in certain cases it is desirable to reduce their content in red or white wines. The principal method to do so is based on the interaction between tannins and proteins or protein-like molecules.

It must be emphasized that the protein-tannin bonds are not ionic, nor covalent bonds, but rather of a hydrogen-bond nature. The hydrogen bonding is multivalent, due to the many sites of hydroxyls and carbonyls in the interacting molecules. The tannin-protein complex may be soluble or insoluble in water, depending on its concentration and on the molar ratio between the two. When tannin is in excess, the complex will probably precipitate, but when protein is in excess, the complex might remain in solution. The practical application of tannin-protein interaction has been in use for a very long time in the leather industry, where hide reacts with tannin to produce the so-called "leather." The reaction between tannins and proteins is between the phenolic hydroxyl as a proton donor and the peptide's keto-imide carbonyl as an acceptor of the hydrogen-bond interaction:

$$- C - NH -$$
$$\parallel$$
$$O - \cdot H - - O$$

H-O — R

The nature of the complex, namely its bond strength, solubility, reversibility, and microbial resistance, depends on the protein (amino acid composition, its molecular weight), the tannin (polymeric size), and the reaction conditions (pH, ethanol, ionic strength). The main phenolic targets for removal from wine are the procyanidins (polymers of flavanol-3 and flavone-diol 3,4) in the molecular weight range of (500 – 3000) daltons, which are mainly responsible for astringency in red wine.

Gelatin: Gelatin is a protein product made from animal tissues (bones and skin). Collagen is the raw material, which after processing (partial hydrolysis, evaporation, drying, and grinding) is sold as a white powder or in sheets. The molecular weight of commercial gelatin for wine fining is in the range of (15,000 – 150,000) dalton and it contains about 85% proteins and up to 2% mineral residue. The rest is water. The gelling power of gelatin (namely, the resistance to deformation of the gel made under specific conditions) is expressed in *bloom number* which ranges in wine gelatin from (80 – 150), with a recommended range to use of (80 – 100). Gelatin is able to absorb water up to ten times its weight. The isoelectric-point of gelatin is in the range of (4.8 – 5.2) above wine pH, and therefore it is positively charged. Gelatin interacts mainly with *polymeric phenols* rather than with low molecular weight phenolic compounds (in other words, with a greater number of hydroxyl sites per molecule in the poly-phenols), and therefore its fining effect depends on the wine age and oxidation state. It is more effective on *aged red* wine than on younger wines. When applied, it can reduce astringency, some bitterness, and polymeric anthocyanins in red wines. It is very useful in reducing the harshness of pressed wine. Some care is needed not to strip the wine flavor and character by over-fining. The range of gelatin addition to red wines is (50 – 100) mg/L. Up to 200 mg/L can be added in certain cases to pressed wine.

The dose range of gelatin in fining white wines is (20 – 50) mg/L. It can clear cloudy wines that are difficult to clear with bentonite. It is also effective in reducing bitter aftertaste in white wines. In order to prevent protein over-fining (and instability), silica-sol (or tannin) is added in combination with gelatin at a quantity range of (10 – 25) ml of kieselsol 30% commercial solution in a hectoliter of white wine. The addition of kieselsol has to be done one day before the gelatin fining. As with other fining agents, laboratory tests must be carried out in order to determine the right quantity before it is applied to the whole batch of wine. The actual quantities of gelatin/kieselsol must also be determined by testing. Racking and filtration of the fined wine must be done after one to two weeks. By such fining, excellent clarification can be achieved. There is no legal limit for gelatin addition in wine.

Gelatin solution is made by adding gelatin powder or sheets to almost boiling water with constant mixing on addition. The solution concentration should be 1% (by adding 10 gram/L, which is 10 mg/cc). The dissolved gelatin is a viscous gel solution which should be added to the wine while it is still hot. (It will solidify on cooling.) On slow addition to wine, thorough mixing is essential to ensure even contact with the entire wine volume. As in red wine, racking and filtration should be done within one to two weeks.

Isinglass: Isinglass is a protein extract made from the swim bladder of certain fish. The commercial product looks like transparent chips. Its molecular weight is about 140,000 and the isoelectric point is in the (5.5 – 5.8) range, well above wine pH. It is used in white wine clarification at very low concentrations (10 – 50) mg/L with excellent results. There is no need for tannin or silica-sol addition in combination with isinglass (as is done with gelatin, see above). The material interacts mainly with the *monomeric phenols* rather than polymeric ones. There is no legal limit for isinglass addition to wine.

The preparation of an isinglass solution is as follows: To 10 liters of cold water add 10 grams of tartaric acid and 4.4 grams of solid potassium-metabisulfite. (This will make SO_2 solution of 250 ppm.) Mix until completely dissolved. Then, while mixing, slowly add 100 grams of isinglass. Stir again a few more times in the following day, whereupon the solution becomes viscous and jelly-like. The solution can stand and be in good condition for days. It contains 10 gr/L of isinglass. With such solution, each cc contains 10 mg of the agent, and for example, if an addition of 20 mg/L is needed to 4000L of wine, 8L of the isinglass solution is required. Racking and filtration can be done in one week.

***Casein*:** Casein is a milky protein which has been used as a wine-fining agent for a very long time. It is prepared from milk as powdered potassium-caseinate, which is water-soluble. In low pH (as in wine) it is not soluble and will precipitate as soon as it is added. It is used in white wines to reduce phenolic bitterness, over-oaked white wine, some off-flavors, and to lighten brown color in oxidized and pinking wines. The casein doses needed are in the range of (50 – 250) mg/L. Milk can also be used for fining, especially low-fat milk, which avoids the introduction of milk-fat into the wine. About (2 – 10) ml/L of milk is the concentration range of milk usage.

To prepare the agent for use, make a 2% (20 g/L) solution of potassium-caseinate in warm water and stir patiently for a long time. Leave overnight and stir again until it becomes completely dissolved. The solution is good for a day or two. Each cc of it contains 20 mg of the agent. When added to wine it should be well mixed. Before racking, bentonite can be added and then racking and filtration can follow.

***Egg-White (albumin)*:** The substance is the white of eggs and contains about 10% proteins (albumin and globulin). This fining agent is solely used for red wine fining and is considered the "best" for softening and polishing red wines. The protein interacts mainly with *higher polymeric* phenols rather than lower ones or monomers. The usual doses are one to two eggs per barrel of red wine (225L). To prepare the egg white solution from fresh eggs, break the eggs and separate the whites from the yolks. Collect all the whites and mix them in salted water solution (0.5 – 0.9) % of table salt. The salt is needed to dissolve the globulin of the egg white, which is soluble only in salted water. Albumin is soluble in pure water. The recommended eggs/water proportion is about 15 eggs per 1L of water. Mix well with a mixer and use it the same day. When the solution is added to the wine (in barrels) it has to be stirred very well. Racking should be done in about a week. Dried powder of egg white is available also, but it is less favored. About 3 grams of dry powder is equivalent to one egg.

PVPP (polyvinylpolypyrrolidone): This is a synthetic polymer whose exact molecular weight is difficult to measure because the substance is not soluble in any common solvent. Its structure is:

$$\left[\begin{array}{c} \underset{\overset{|}{\text{N}}}{\bigcirc} =O \\ \text{-- CH -- CH}_2 \text{ --} \end{array} \right]_n$$

PVPP is a higher molecular weight polymer of PVP (polyvinylpyrrolidone).

The commercial substance is sold in granular form with mean particle size of about 100μ. Being polyamide, PVPP interacts with phenolic compounds as other agents do. But unlike most of them, it reacts specifically with *low phenolic* such as monomers and dimers (e.g., catechin and anthocyanin). The interaction takes place between the PVPP carbonyl group and the phenolic hydroxyl. Because the PVPP is insoluble, the phenolic molecule adsorbs on its surface and precipitates out of the solution. In white wine, PVPP is useful for reducing brown color and pinking and as a preventative measure when such risk is forecast. It will contribute to color stability in sensitive blush wines. When used for color reduction in white wine, a combination with carbon is more effective in many cases. It helps also to settle down the carbon particles. In many cases PVPP can also reduce certain off-flavors and bitterness. Performing a trial to see if it works in a specific case might be beneficial. The dose range in white wine is (100 – 700) mg/L. In red wines it is less frequently used, but it can reduce bitterness and brighten the color. The dose range in red wines is about (100 – 200) mg/L. Addition of PVPP can be done at any stage of production, from must to pre-bottling. PVPP is a gentle fining agent which almost does not strip wine aroma as other agents do. PVPP can be added as powder directly into the tank while stirring. Mixing the powder with water or wine in (5 – 10) % slurry prior to addition to the tank is also helpful. It interacts very fast and settles within a few days. The legal limit of PVPP in the US is 800 mg/L.

3. Non-Specific Fining

Carbon: Activated carbon is made of very small particles with an extremely high surface area which ranges about (500 – 1000) square-meter/gram. The carbon particles' surface is capable of adsorbing both gases and liquids by physical adsorption of *Van Der-Waals* attraction forces. The characteristics of the absorption, that is, the concentration of the adsorbed substance per carbon weight unit as a function of the adsorbed substance concentration in the medium under study, can be described by either *Langmuir* or *Freundlich* equations. The formulae work in either medium, liquid or gas phase.

In wine, carbon can be used for removing off-flavors and odors of various kinds and to decrease browning or pinking in white wines. The carbon type most suitable for decolorizing is marked as *KBB*, and the deodorizing type is marked as *AAA*. As mentioned above, it works well in combination with PVPP in both tasks. The adsorption on the carbon surface is very fast, so the results are seen immediately. Due to its absorption capability, carbon will also absorb aroma and flavor components, reducing their concentration in the wine, so care should be taken when carbon is used, and moreover, it should be used only if no other means to solve the problems is possible! (For example, $CuSO_4$ can reduce the hydrogen sulfide or mercaptan very successfully instead of carbon.)

The range of usage is between (1 – 5) gr/HL or (10 – 50) mg/L for color treatment and (5 – 25) gr/HL or (50 – 250) mg/L for removal of off-odors. The exact quantity required should be determined in the laboratory prior to its addition to the wine. In the laboratory test, the samples should be mixed with carbon, filtered after an hour, and checked for improvements of color or smell. When added to the bulk wine, at temperature ranges of (15 – 25) °C the black carbon powder is administered directly and well mixed. It will settle in a few days. The use of carbon was not restricted in the US until recently. It is now limited to 3.0 gr/L, which is far above the practical range.

A summary of the various fining agents, their use, dose range, and application methods is given in the following table:

Agent	Use for	Dose range	Preparation and use
Bentonite	* Clarification - white wine. * Stabilization - white wine against protein haze formation.	(0.1 – 1.5) gr/L Less bentonite is needed in low pH wines.	Add slowly 50 gr/L of bentonite (5%) to boiling water while stirring. Let stand for a day with occasional mixing. It will become viscous slurry solution. The solution is good for a few days. 1 cc of slurry contain 50 mg of bentonite. Do not fine very cold wine (below 15°C).
Gelatin	* Clarification - white wine. *Tannin and bitterness reduction in white wine. * Tannin and astringency reduction in red wine. * Tannin and astringency reduction in pressed red wine.	(10 – 25) mg/L (25 – 50) mg/L (50 – 100) mg/L (50 – 200) mg/L	Add 10g/L of gelatin to nearly boiling water and stir well to form a gel solution of 1%. Add slowly to wine while it is still warm, with constant stirring. Rack off after one-two weeks. Solution contains 10 mg/cc of gelatin.
Kieselsol	In combination with gelatin for white wine clarification and bitterness reduction.	5 times (in ml) of gelatin added (in grams/HL), e.g. 50 ml/HL for 10 gr/HL of gelatin added.	The commercial agent is a 30% solution prepared for direct use. Addition one day prior to gelatin addition is recommended.
Isinglass	* Clarification of white wine to brilliancy.	(10 – 50) mg/L	To 10 L of cold water add 10 gram tartaric-acid + 4.4 gram $K_2S_2O_5$. Mix until dissolved. Add 100 gram isinglass mix and let stand one day with occasional mixing. Discard after one week. The slurry solution contains 10 mg/cc of isinglass.

Continued next page

Agent	Use for	Dose range	Preparation and use
Casein	* Bitterness reduction in white wine. * Over- oak reduction in white wine. * Browning and pinking reduction in white wine.	(50 – 250) mg/L of casein or (2 – 10) mg/L of low-fat milk.	Add 20g/L of casein-salt to warm water and stir well. Leave overnight and repeat stirring. Good for use for one to two days.
Egg white	Softening and polishing red wine.	(1 - 2) egg whites per barrel (225L).	Collect egg whites from fresh eggs. Dissolve in (0.5 - 0.9) % salty water. Mix with a mixer and use freshly made.
PVPP	* Color reduction in white wines (browning and pinking). * Stabilizing white and Blush wines against browning. * Reduction bitterness in white and red wines.	(100 – 700) mg/L in white wines. (100 – 200) mg/L in red wines	Direct addition to wine or mixing with wine to (5 - 10) % slurry prior to addition to tank.
Carbon	* Deodorizing off-odors * Decolorizing browning or pinking in white wine.	(50 – 250) mg/L	Direct addition to wine. Settles in few days. PVPP helps to settle faster.

D. Filtration

Cloudy wine will clarify if left to stand for a long time. However, even a wine that looks clear after a long time of settling contains enormous amounts of small particles in the microns (10^{-3} mm) range, including yeast cells, bacteria, pigment particles, proteins, fining particles, tartrate crystals, pulp, and other components. The purpose of filtering is to remove most of these particles from the wine. Filtration is done by two basic mechanisms. The first is by absorption of the particles on and in the surface of the filter texture, by electrical or cohesion forces. These filters are made of fiber pads or a composite of fibers and mineral particles. The filters are therefore called *pad filters*. The wine particles are trapped in the filters in the whole volume of the pads' surface area. The second mechanism is by size control of small pores which prevent any particle bigger than the filter pores from getting through. These filters are called *membrane filters* and are used mainly for very fine and sterile filtration.

1. Pad Filters

Pad filters are made of fabric or paper pads which are mounted tightly in a stainless steel frame (see figure). The mechanical set, which is called a plate filter, contains a series of plates that hold the pads between them (one plate, one pad). The common dimensions of the commercial pads are squares of 20x20 or 40x40 centimeters. The pads are not identical on both sides. One side is rough, facing the inlet wine flow, and the other side is smooth, facing the outlet wine flow. The pads are packed alternately between the plate filters (see figure).

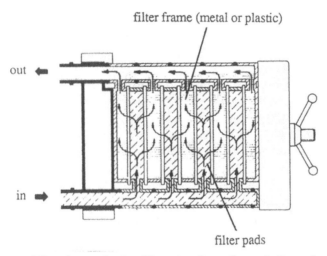

Filter frame device. The wine flows through the pads,
leaving the solid particles inside and on the surface of the pads.

When placing the pads in the frame, one should be aware of maintaining the proper order of the pads, to avoid having tiny particles and fibers from the pads enter the wine. The pads are arranged so that the wine flows in parallel through the filter (see the figure).

The fineness of the filter texture is responsible for two parameters: the distribution of the filtered particle sizes that can pass through the filter, and the resistance of the filter to through-flow. The particle sizes in wine which have to be filtered range between sub-micron sizes (bacteria), up to dozens microns (lees, fining particles and tartrate crystals). Commercial filters are numbered in a series according to the fineness of the pad filter. The higher the number, the finer the filter. Sterile filters are marked by letters (such as EK, EKS etc.).

When wine flows through a filter, the rate of filtration (volume of filtrate per time unit) is directly proportional to:

(1) The pressure difference between the input and output of the filter.

(2) The filter area.

(3) Inversely proportional to the viscosity of the liquid.

(4) Inversely proportional to a factor called "resistance to filtration." This term depends on two sub-factors:

(a) The filter's resistance to the flow of clear water, an inherent characteristic of the filter texture (pore size and packaging tightness). This is the "internal resistance" of the filter.

(b) The filtered particles that were left inside and on the filter when the filtration is done. This resistance is constantly growing as the filtering is continuing. The total *filtering resistance* is the sum of the constant internal resistance and the accumulating one. The above parameters can be expressed in a formal relationship:

$$\mathbf{dV/dt} \; = \; \frac{\mathbf{dP \times A}}{\gamma \times \mathbf{R}}$$

where: dV/dt – is the volume of filtered wine per unit time (filtration rate).

dP – is the pressure difference across the filter.

A – is the total filter area (pad's area x number of pads).

γ – wine viscosity.

R – filtering resistance.

The flow rate of typical commercial filters is given in the following table:

Filter mark	20	30	40	50	60	70	EK	EKS
Liter/hr/m^2	6600	6000	4600	3750	3400	2500	1100	750

Flow rate of water through filter pad at pressure difference of 10 p.s.i.
The flow rates are in liters per hour, per one square meter area of the pads.
(One square meter area of 40x40 cm pads is equal to ~ 6 pads.)

After the filter plate is set up and the pump has been connected, it is very advisable to flush the whole filter setting with 0.5% tartaric acid or citric acid for a couple of minutes (in a closed cycle) in order to avoid introducing the "papery" flavor of the filter pads into the wine. The filter pads, the hoses, and pump should be emptied from the solution after the flushing, before connecting to the wine tank. The filtered particles are accumulated on and in the pads, causing its flow resistance to become higher, and as a result, pressure is built up across the filter surface between the input and output of the filter. The increasing resistance gradually reduces the flow rate on one hand, and increases the built-up pressure on the other hand. The high pressure can cause a break in the filter texture, causing the filter to lose its filtering capability. At the same time, the flow rate becomes tediously low. Therefore, the differential pressure should not be higher than about (2 – 3) bars (30 – 40 p.s.i). When it reaches that pressure, it is recommended to replace the filter pads.

Another important factor that must be taken into consideration is the exposure of the wine to air during filtration. To minimize oxidation, especially in white wine, the addition of sulfur dioxide to the wine before filtration is highly recommended. For the first filtration, which is done after bentonite and cold stabilization, a suitable option is to add SO_2 before the bentonite addition and cold stabilization, and then filter the wine. (For details see the previous section on stabilization.)

Some wineries use filter aids (diatomaceous-earth or kieselguhr), which are very small mineral particles with a very large surface area. They do it to reduce clogging of the filter pads. The powder of diatomaceous-earth (DE) is mixed with the wine just before filtering. When the wine is flowing through the filter, it leaves the DE particles together with the wine particles to build up layers on the pads. The mixed multi-layers of DE ease the flow for some time. In certain filter devices the DE is the filter itself (replacing the use of paper pads), after it has been built up on a metallic screen as a filtering layer. The layer is very fragile and is kept in place by the continuous flowing pressure. In other filter devices, the filter-aid is introduced into the wine by a dosing unit mounted at the input of the wine entrance. The amount of filter aid which is added to the wine for that purpose is in the range of (50 – 100) gr/HL, depending on the quality of lees particles in the wine.

2. Membrane Filters

These filters are made of synthetic polymers (cellulose esters) that have uniform tiny holes at the micron range. The membrane filter is a *surface filter*, in contrast to the pad filter which is a *depth filter*. In a membrane filter the flowing particles are blocked on the *membrane surface* by the size of the holes which specify the filter mesh characteristic. The common sizes in membrane filters are (1.2, 0.65, 0.45) microns. The bigger size will block most of the yeast cells and the finest size will block most spoilage bacteria. The membrane filter is assembled as a plastic cartridge mounted within a vertical stainless steel tube. The wine flows into the cartridge, leaving the microscopic particles on its outer surface. During operation, the accumulation of particles on the surface increases the resistance and causes pressure to build up. As mentioned above, the pressure builds up and the flow rate decreases. At high pressure, the filter may be permanently damaged, losing its sub-micron filtering capability. The maximum working pressure is about 3 bars (~ 40 psi). When reaching this pressure, filtration should be stopped, and the filter cleaned. In order to clean the membrane filter, water flow from the opposite direction (from output to input) will remove the particles which were left on the membrane's surface. The membrane filter is generally used only before bottling the wine, directly into the bottle. This kind of filter can be used *only with clear* wine having a low concentration of particles. Otherwise, it will immediately be blocked. Therefore, a sterile membrane filter should be used only after pad filtering has been done.

At bottling, because the sterile filtering is the major measure taken for biological stability, and in sweet or off-dry wines it is sometimes the only assurance against re-fermentation, it is most important to check the filter before each bottling day (and at the end of the day too!). The checking is done by the so-called "bubble point" test. To run the test, water flows first through the filter to fill it with water. Then a nitrogen line from a cylinder is connected to the inlet of the filter, and the outlet is connected by a thin plastic tube to an open vessel containing water. The main nitrogen valve is then opened, and by using the regulating valve, the pressure is slowly increased until bubbles of nitrogen appear at the filter outlet (in the water vessel). This pressure is needed to force the bubbles of gas to act against the water surface-tension across the filter's holes. The smaller the radius of the hole, the greater is the pressure needed. If there is a break in the filter texture, the bubble pressure would be lower than the normal test pressure. For 0.45µ cartridge (the common filter for white wine bottling) the bubble point pressure should be between (25 – 40) psi. If the bubble point appears at lower pressure, the cartridge should be replaced.

3. Mechanical Separators

There are two mechanical devices which can be used to clear turbid wine:

One is a continuous diatomaceous-earth filter (kieselguhr) which is based on building a layer of small irregular DE particles like a filter texture on a permanent support, preserving these filtering layers with a constant flow of wine. The advantages of the diatomaceous-earth (DE) continuous machine are: (1) high flow rate; (2) very high filtering capacity before the filter gets stuck, especially when the solid particle concentration is high (in must or young wines); and (3) economic advantage compared to pad sheet filters (DE is much less expensive than filter pads). The disadvantages are: (1) Complicated operation; (2) long preparation time before starting the filtering, about (1- 2) hours; and (3) operational error may lead to the necessity of starting the whole preparation again from the beginning. The amount of DE that is needed for building the "cake" before filtering is about $(1.5 - 2.0)$ Kg/m^2. During filtering, the DE is continuously fed into the wine by a dozer from the DE container in the machine. There are different particle sizes composing the DE, suitable for the different filtrate particles. For must or young wines, Celite 535, 545 can be used. For clean wines, Celite 501, 503 are preferred. During preparations, before filtering the wine, the "cake" should be washed (in a closed cycle) with 0.5% tartaric acid or citric acid (to remove the "papery" taste), and then with clear water. The initial differential pressure before starting the filtering is about 1 bar. The working pressure can rise up to $(6 - 7)$ bars, and then the filtering should be stopped; the machine has to be cleaned and prepared from the beginning. The operation of this filter device is quite complicated and sensitive, and in our opinion it is suitable for a large quantity of wines, namely, for medium size wineries and up. There is a bigger version of this principle where the filtering layer is built on a big rotating drum. This device is suitable for wineries with huge quantities of wine.

The second machine for clarifying must or wine is the centrifuge, which can clear turbid wine easily and quite satisfactorily. But it is expensive and perhaps not economical for a small-scale winery.

One final remark on filtration pumping: When doing filtration, it is better to use a *continuous pressure* pump (such as a centrifugal pump or oval one or peristaltic pump) for the filtering operation, rather than a pulsed pressure pump (such as a piston pump or air membrane pump), which generates cyclic pressure on the filter rather than a continuous smooth one.

E. Blending

Blending of wine is done in order to achieve different goals:

(1) Overcoming certain deficiencies or defects.

(2) Balancing the wine.

(3) Enhancing complexity.

Blending can be done at any time along the way of winemaking, from must blending before fermentation and at any wine-processing stage afterwards. The corrections made by blending two wines together should enhance both of their respective characteristics; one may be over-endowed with a certain character and the other may suffer from a deficiency of it. By blending them, both wines will improve in that specific character. However, there are many other parameters involved in wine quality, and all of them have to be considered. The "art" of blending is not a simple one. Blending can be done with different varieties, different vintages of the same variety, different vineyards or locations of the same variety, and different lots of the same vintage (tanks or barrels).

Each country has its special rules concerning blending and labeling of wines. In California, for example, in order to state a varietal name on the label, it is mandatory to blend at least 75% of that specific variety in the wine. If the blending is such that there is less than 75% of any variety, the wine has to be labeled as generic wine. This does not mean, however, that a generic wine cannot be an excellent wine. As for the region of origin (appellation), the approved American Viticulture Appellation (AVA) requires that 85% of the grapes must come from the indicated region in order to be stated on the label. The regions do not necessarily coincide with geographic boundaries.

For vintage statement, 95% of the wine composition must be from the year that is printed on the label. The alcohol content in table wine should be in the range of (7 – 14) %. The winemaker should be aware of all specific regional rules when blending his wines. The same principles of specific regulations are practiced in other countries as well.

In varietal blending there are certain conventions regarding which varieties are suitable to blend. For example, Cabernet Franc and Merlot are commonly blended with Cabernet Sauvignon, or Semillon with Sauvignon Blanc. French Colombard is a good blend with any fruity white wine because of its high acidity content and floral aroma. In certain cases of generic wines, there may be up to a dozen varietal blends with excellent balance and quality. It may contain red and white varieties as well (Tuscany, Provence).

In regular cellar operations, it happens quite frequently that certain lots of the same vintage develop differently, better or worse, than the average of that vintage. In such cases, the question is whether to separate these lots (if better, as special reserve, or if worse, to sell it as bulk) or to blend them together to get the best from what is at hand. No formula or recommendation can be made. The decisions are ad hoc. Also, if a press run is separated, then special attention must be paid at blending, because it may contain high concentration of tannins, high pH, volatile acidity, and a flat taste.

The parameters that can be easily corrected by blending are: acidity, pH, alcohol, color, tannin, varietal aroma, freshness and fruitiness, oak flavor, volatile acidity, residual sugar, bitterness, and off-flavor. Some of these parameters are in excess (defining the wine as having some noticeable flaw) or in deficiency (causing the wine to be considered unbalanced, of lower quality). Blending may bring these parameters to the desired level, or at least make them tolerable. Some of the parameters are quantitatively measurable (like acid, residual sugar, alcohol, volatile acidity, color) and their concentrations are linearly related. Other parameters, such as aroma, flavor, off-flavor, can be appreciated mainly by organoleptic judgment.

For measurable parameters, the volume ratios of the wines to be blended, according to the desired concentrations of the parameter in question, are given by the simple formula

$$(7) \quad \boxed{P_1 + XP_2 = P_b(1 + X)}$$

where: P_1 – concentration of the actual parameter in wine No. 1

P_2 – concentration of the actual parameter in wine No. 2

P_b – concentration of the desired parameter in the blended wine

X – ratio of wine No. 2 volume, to wine No. 1 volume ; $(X = V_2 /V_1)$

For example, wine No.1 with 8 g/L residual sugar is blended with wine No.2 which has 38 g/L sugar, to obtain a designed blend with 18g/L of residual sugar. Then 8 + 38 X = 18 (1 + X), with X = 0.5; thus, by taking two volumes of wine No.1 and one volume of wine No. 2, the blend will contain 18g/L of residual sugar.

For those parameters that are not directly measurable, a process of trial and error must be carried out to find the best result. All blending tests must be tried first in the laboratory. The aid of close colleagues is highly recommended. When the attempted formulation has been concluded, a small sample of five gallons can be blended and left to "marry." The "marriage" takes a week or two, and then the wine is tasted, and if satisfactory, the whole batch of wine can be blended accordingly. If there is some doubt about the exact wines ratio, the five-gallon blend can be retested with some small variations, and after a short "marriage" period a final decision can be made as to which is the best blend.

When the wines are blended at advanced stage of processing (before bottling), stability problems might sometimes arise, especially precipitation. For this reason it is best to blend early, before stabilization takes place. On the other hand, only when the wine is in advanced maturity stage, can one evaluate the full character and quality of the wine. It is our opinion that after blending, the storage temperature of the blend should be (5 – 10) °C. This temperature should be maintained for couple of weeks to check for instability. If there is any, for whatever reason, care should be taken before bottling.

F. Maintenance

1. Sanitation

Besides its aesthetic value, winery sanitation is most important in preventing and minimizing bacterial and fungal spoilage of the wine. Although wine is less susceptible to harmful spoilage microorganism than other food products, it is still sensitive to a number of microorganism infections which may affect its quality considerably. Winery sanitation includes cleaning and sanitizing all the equipment in the winery that comes in contact with the wine, from crushing to bottling. Of all the many materials and cleaners on the market, we choose to emphasize here those that are most efficient and most used in the wine industry.

Water: Can be used cold, hot, and as steam. Cold water cleans by dissolving and physically removing dirt from the surface. Hot water is much more effective than cold water. At 80°C water has a double effect, namely, physical cleaning and sterilizing at the same time. Three major devices are in use for applying water cleaning in the winery:

(**1**) *Water guns* use high water pressure. Spraying water with a water gun is the most common and basic technique for cleaning in the winery. During harvest time, at the end of any working day, all the machines that have been used (hopper, press, drainer, destemmer/crusher, hoses, pumps, tanks) should be washed with water to remove all residues.

The hot water machine is the most useful piece of equipment in the winery. It is connected to the water line and can heat the water instantly up to about 80°C and eject it at very high pressure through a gun. The cleaning ability of this machine for any surface (metal, cement, plastic, and wood) is most effective. It also very effectively conserves water. The internal parts of equipment like the pumps and hoses can be cleaned with hot water circulating in a closed cycle.

(**2**) *Sprinkler balls* can be operated by regular pump pressure or by the winery water system pressure, and can be used inside containers like tanks and barrels, to spray the water uniformly on the container walls.

(**3**) *Hot water* can be used as a sanitizing agent against almost any microorganism. Its effectiveness depends on the temperature and the time of application. At 80°C and 10 minutes of hot water contact, almost any bacteria and fungi (but not all spores) would be killed. For example, the filling machine in the bottling line has to be sanitized at the beginning (and end) of every work day by passing through hot water at 80 °C for (10 – 15) minutes.

Alkaline solutions: The strongest agent in this category is sodium or potassium-hydroxide. Its main cleaning function is to soap oily dirt to emulsify it, so that it can easily be removed by water stream. Milder than sodium-hydroxide but still alkaline in its activity is sodium-carbonate (also called soda ash). Alkaline solutions can also be used to dissolve the hard layer of deposited tartrate on the inside walls of tanks and barrels, or any equipment that has come in contact with cooled wine or must for an extended period of time. Alkaline solutions are used at a concentration of about (0.2 – 0.5) % in water. When the solid hydroxide is dissolved in water, it releases heat, and care should be taken whenever this agent is used. Alkaline solutions are skin-irritating and very dangerous to the eyes. Glasses and rubber gloves are the minimum safety measures when working with alkaline.

When preparing alkaline solutions, a general practice is to prepare the solution in a small container of (100 – 200) liters, and to apply the solution to the equipment (such as tanks) by spraying it with a sprinkler ball in a closed cycle. The cycle proceeds as follows: solution is pumped from an open container through a hose into the tank and then back into the container. On cleaning, the tank's openings should be closed for safety reasons. About (10 – 20) minutes of application is usually enough to remove any sediment from the tank's walls. To clean small equipment, immerse it in the container for a couple of minutes. Barrels are best cleaned (if necessary) with a less potent alkaline, soda-ash. They should be filled with the solution for (15 – 30) minutes and then emptied and washed.

The residue of alkaline on the surface after cleaning is difficult to remove by water. It stays on the surface in a very thin layer. Therefore, after any application of alkaline, the surface should be rinsed with water and washed with 0.1% citric-acid or tartaric-acid solution (which interacts with the alkaline surface residue to form easily dissolved salt) and then finally rinsed with clean water to wash away the acid.

Chlorine solutions: These are based on Sodium (NaClO) or Calcium $Ca(ClO)_2$ Hypochlorite which in solution exist in two forms: hypochlorite anion (ClO^-) and hypochlorous acid (HClO). It is produced by reaction of chlorine in water to form a mixture of hydrochloric-acid and hypochlorous-acid:

$$Cl_2 + H_2O \quad <====> \quad HClO + HCl$$

In aqueous solution hypochlorous-acid partially dissociates into *hypochlorite* ion:

$$HClO \quad <====> \quad ClO^- + H^+$$; and their relative concentrations depend on the solution pH.

At low pH when the undissociated acid (HClO) is predominant, the solution acts as an excellent sanitizing agent against most microorganisms. Its action is based on its ability to chlorinate the cell's enzymes of microorganisms after penetrating into its cell wall.

At high pH, when the hypochlorite-anion is the predominant form, it has a powerful oxidation potential which can bleach organic stains and cause them to emulsify and be removed by excess water.

The commercial reagents are sold as *bleaching* agents (hypochlorite + alkaline) or as *chlorinating* agent (hypochlorite + acid). The former is usually sold as powder and the latter as liquid. The powder form has an effective concentration when it is dissolved at about 5 gr/L, and the liquid quantity depends on its initial commercial concentration. In any case the exact producer's instructions should be followed for good cleaning effect and for safety awareness. Here again, as with the alkaline solutions mentioned above, if the bleaching powder is used (alkaline agent), it has to be followed by washing with 0.1% citric or tartaric acid and then with water. *Again, safety measures should be taken as stated above when dealing with these agents!*

Iodine: An excellent sanitizer. It consists of a solution of iodine in water (plus potassium iodine to increase solubility), and surfactant materials to increase its surface contact. The solution's pH is set to about 4 by phosphoric-acid. This solution is active against microorganisms due to its free iodine. The brown color of the solution makes it easier to follow when rinsing it with clear water after use. The effective solution concentration is about 25 mg/L, but the exact quantity should be read on the producer's instructions.

2. Storage

Wine can be stored in different kinds of containers: wood, concrete, iron, plastic, and stainless steel. When concrete and iron were used, the interiors of the containers were usually covered with glaze, ceramics, or polymeric paint. In modern wineries, wood, concrete, and iron containers are rarely used any more; stainless steel has replaced them for ease of maintenance and its inert surface. Plastic containers, which are much cheaper than stainless-steel, are also rarely used in the wine industry, probably because in the long run their amortization is higher than stainless steel containers. They are also harder to clean and are not as inert as stainless steel. Oak containers (barrels and casks) will be discussed in a separate chapter.

The cleanliness of containers (or tanks) is a basic requirement in cellar operation. This means that immediately after racking, the tank should be washed, brushed if necessary, and be left empty and clean for its next filling. Before filling any tank with wine, it is highly recommended to recheck the tank through its opening.

When storing white wine during vinification stages and operations, the main objective should be to minimize air (oxygen) contact. The reasons are twofold: preventing oxidation and preventing acetobacter infection. In red wine, only the second reason is of major importance. To minimize oxygen in white wine storage, it is necessary to keep the tank as full as possible, and also to fill the space above the wine with an inert gas such as carbon-dioxide or nitrogen. The gas lies on the wine surface as an inert blanket, protecting it from contact with oxygen. Carbon-dioxide is heavier than air and will stay on top of the wine, but it is highly soluble in the liquid. Nitrogen is much less soluble, but it is lighter than air and may diffuse out. When using either gas, the top of the tank should be checked and refilled from time to time if necessary. The cover on top of the tank should be closed, and an air-locked valve should be mounted on it to minimize gas diffusion and to prevent any pressure differences that can be formed between the inside and the outside of the tank (by temperature variations).

As for the wine temperatures during storage: for white wine the preferred range is (8 – 12) °C and for red wines (20 - 25) °C.

BARREL AGING

Chapter v: Barrel Aging

Wooden cooperage adds an *extra cost* to wine production due to the high invest-
ment price, increased labor, and the need flor a much larger storage space, as well
as losses caused by wine evaporation from the barrels. In spite of all of these extra
expenses, certain wines (red and white) have to spend some period of time in small
oak cooperage in order to finish their maturation before being bottled.

It is generally accepted that one of the most important parameters in red wine
quality is the balance between the varietal aroma, the aging bouquet, and the oak
character. The last two angles of this quality triangle are achieved by aging the wine
in oak cooperage. Wooden containers are not easy to maintain because of leakage
problems, sanitation difficulties, difficulties in temperature control, and problems
of maintenance when not in use. They are also expensive. So why do we use them?
The answer can be found by understanding the goals of aging wine in oak barrels:

(1) *Slow Oxidation* of the wine, which enhances its bouquet and softens its
tannin by slow polymerization of its phenolic compounds. This function is fulfilled
by the periodic topping operation, which is needed because of the evaporation of
the wine through the wooden surface.

(2) *Adding Oak Phenolic* into the wine, by extraction from the inner surface of
the barrel. These compounds are most compatible with the wine aroma component,
and together they expand the complexity of the wine's bouquet.

A. Cooperage

1. Barrel Making

The basic details regarding the origin, making, characteristics, and qualities of barrels will be presented in this section.

Origin: The factors that people considered many years ago when they had to choose the tree species to use for wine storage were:

(1) Its wood strength.

(2) Large trunk diameter for economical production.

(3) Straight and clean wood with no defects or knots to enable clean-cut staves with low chance of leaking.

(4) Good flexibility.

(5) Absence of negative undesirable extraction components which might be transmitted from the wood into the wine.

Redwood fulfills most of these demands, but it contains too high a level of extractable phenols, which may damage wine stored over long periods. Red oak is too porous for reasonable wine storage. Other trees were found to have an objectionable flavor or color which affects wine stored in their barrels. No less important is to have a *positive desirable* flavor extraction potential which is compatible with wine aroma. The ultimate wood which meets all of the above qualifications is *white oak*. It has all the mechanical factors as well as the desired aroma, which is the most important factor.

A cross section of a tree is shown in the following figure:

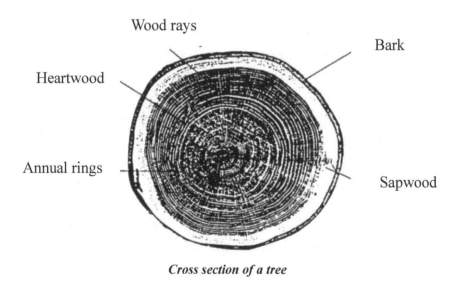

Wood rays

Bark

Heartwood

Annual rings

Sapwood

Cross section of a tree

Its features are (from outside to the center): *Bark,* which has two parts: the outer corky dead one, and the inner living part. Next to the bark is a light-colored section called the *sapwood,* which is the wood tissue that carries sap from the roots to the leaves. The sapwood is softer and very permeable, and therefore cannot be used for barrel making. The next inner section, which is darker and harder, is called *heartwood,* and its function is to give the tree its mechanical strength. Only the heartwood is suitable (after drying and seasoning) as cooperage raw material.

The oak genus *Quercus* has about 300 known species. There are basically two groups of oak trees, *red* and *white* oaks, and only some of the white oaks are suitable for cooperage. The main European oaks are *Quercus-robur* and *Quercus-sessilis.* *Quercus-robur* is also called *Q.-pedunculata,* and *Quercus-sessilis* is called *Q.-sessiliflora* or *Q.-petraea.* The two are considered different varieties of the same species. European oaks are distributed all over Europe, mainly in France, Italy, Portugal, Hungary, Slovakia, and Russia.

The main American oak is the *Quercus-alba* species, grown mainly in Kentucky, Missouri, Kansas, Oklahoma, Arkansas, and Texas. Besides the species difference, different geographical regions produce variable kinds of woods, which are different in their physical and chemical characteristics. European oak is definitely different in its flavor from the American oak, and between the two European varieties, *Quercus-sessilis* is more tannic and contains more extractable solids than *Quercus-robur.* These two varieties are not always separated, and in many forests they grow side by side. Because France is one of the major producers of wine (and brandy) barrels, we will pay more attention to French oak. The main regions where oak is grown are Limousin, Tronçais, Allier, Nevers, and Vosges. It is very difficult to assure the origin of the oak, because oak logs pass through so many hands and are mixed in the wood mill by those who cut them into staves. Instead, it is common to separate the staves according to the coarseness of their grain (fine, medium, and coarse). Limousin is considered to have a very coarse grain, Nevers a medium grain, and Tronçais, Allier, and Vosges a fine grain. Analytical assignment of the composition profile of eleven major components in oakwood was recently studied, with a provisional conclusion that American and French oak can definitely be distinguished from each other, whereas the differences among the French regions themselves are not so clear. More studies are needed to verify and establish the differences between French and American oaks.

Barrel Production and Characteristics

Before describing its production, let us be familiar with some barrel terms shown in the following figure:

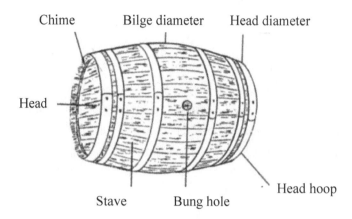

First, the trunk is sawn into logs of the desired stave length of about (90 – 95) cm for the 225L barrels. Each log is then split radially (with hydraulic wedge or an axe) at the cross section of the log, so that sections are split along the natural rays of the wood. This is preferred to the other option of *sawing* along the length of the log, which does not follow the natural wood rays. Splitting the wood prevents leaking from the tiny channels along the rays. Splitting is much more expensive than sawing because it is a hand operation, and also produces more waste material than exact cutting with a saw.

The sapwood is cut off from the split quarts, and only the heartwood is used for making the staves. The quarts (split) are then cut (now by sawing) to staves, so that from each split several staves can be produced.

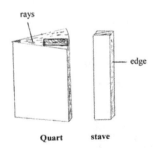

The wood rays are then parallel to the stave's surface (perpendicular to the barrel radius), which otherwise would be along the radial direction of the barrel. Because the wood rays run along the staves, in many cases when the barrel is leaking, the leak is found on the chime of the barrel and not along its side (the diffusion takes place all the way out through the stave length). If there is some leakage, it is very much less than if it leaked straight along the stave width, which is only (22 – 28) mm thick. Furthermore, being parallel to the staves, the rays contribute to their flexibility and bendability when heated to form the barrel.

After cutting, the staves have to be dried, either in the open air or in a kiln (mostly for whiskey barrels). Open-air drying (exposure of the staves to rain and sun) may take from eighteen months up to three years. The desired moisture content of the wood, when seasoning is completed, is (16 – 18) % water. If the wood is not dry when the barrels are made, they will shrink later and will not hold wine without leakage. The seasoned staves are then shaved to the exact tapered shape so as to fit in place when assembled into a barrel. The staves are assembled together (held with six to eight hoops) and bent with the aid of open fire and mechanical tools.

After the staves have been assembled and held with temporary hoops, most barrels are toasted in an open fire before the following stage of production (see next paragraph on barrel toasting). The temporary hoops can now be replaced by (6 – 8) galvanized iron hoops. Heads are placed at both sides, a bung is drilled in the bilge of the barrel, and a final polishing of the surface is given to the newborn barrel. Before releasing, a final leakage test is done for each barrel. The check is made by inserting a few gallons of water into the barrel and applying pressure through the bung, rolling the barrel on the floor and checking the outer surface for signs of leakage. If there isn't any, the barrel is certified for carrying wine, and it is ready for shipping.

There are certain dimension ratios which are common in building barrels.

Assign the diameter at the head as **a** ; diameter at the bilge as **b** ; and the barrel length **c** ; The ratio **b/a** , and **c/b** are **1.3 +/- 0.1**.

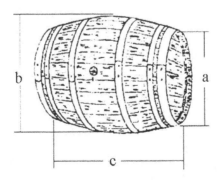

The extraction rate of the barrel's phenolic into the wine depends on the inner *surface to volume* ratio of the barrel. The following figure shows this ratio as a function of the barrel's volume in cylindrical barrels with the above relations of diameter and length:

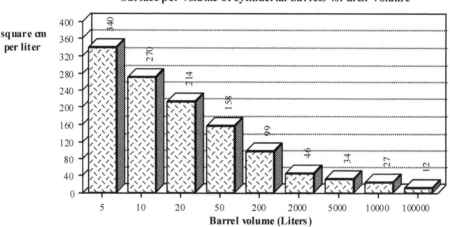

The inside surface of the most common barrel of 225L is about 2m^2, which supplies about 90 cm^2 of wood surface per liter of wine.

The surface/volume ratio between two different barrel-volumes (keeping the head diameter/length ratio of 1.3) is related very closely according to the empirical formula

$$[S/V]_2 \, / \, [S/V]_1 = (2.15)\log V_2/V_1$$

where: **S/V** is the surface to volume ratio of barrel **2** to barrel **1**;
   V_1 and V_2 are the volumes of the two different barrels.

For example, for two barrels $V_1 = 10$ L and $V_2 = 10,000$ L, the S/V ratio is: $(2.15)^{\log 10000/10} = (2.15)^{\log 1000} = (2.15)^3 = 9.94$, as can be verified from the above figure (270 cm^2/L and 27 cm^2/L respectively), where the ratio is 10.

The decision about which volume is most suitable for wine aging has to take into account the storage time needed at each volume, the handling and maintenance effort for each barrel size, and the basic cost of wood per wine volume. Practically, the most common barrel size in winemaking is the French style 225L (60 gallon) called *barrique*, and the American style 190L (50 gallon). Other common sizes are 300L, 350L, 400L, and 500L.

Barrels are built in two major shapes, namely *Bordeaux* and *Burgundy* styles.

The *Burgundy* style is somewhat wider and shorter than the *Bordeaux* one. Each style has two types: the regular type which is called *Bordeaux-chateaux* and *Burgundy-tradition*, and the massive type (for transportation), which is called *Bordeaux-export* and *Burgundy-export*.

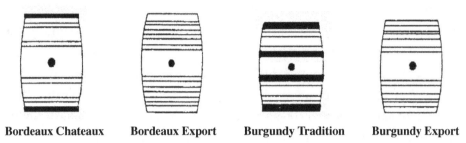

Bordeaux Chateaux   **Bordeaux Export**   **Burgundy Tradition**   **Burgundy Export**

In accordance with this, the dimensions of a typical Bordeaux barrel (±1cm) are: the bilge diameter 70 cm, head diameter 57 cm, and length 94 cm. A typical Burgundy barrel dimensions are (in the same order): 73, 62, and 87 cm. The stave thickness of a Chateaux barrel is about 22 mm, whereas an export one is about 27 mm. An empty barrel weighs about 45 kg when dry.

Barrel Toasting: It is very common to toast the inside surface of the barrel before the heads are placed in position. The toasting is done with an open fire of small oak cuts at three levels of toasting: *light, medium,* and *heavy*. It is most important that the toasting be done gently, at even intensity over the whole surface, so as not to burn it to charcoal. *Light* toasting is done just on the surface, with no penetration to the inside of the wood. *Medium* toasting is more intense and penetrates into the wood to about 2 mm from the surface, while *heavy* toasting goes as far as (3 – 4) mm deeper.

Toasting levels of oak barrels

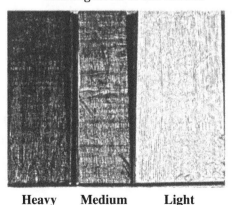

Heavy   **Medium**   **Light**

In whisky (or bourbon) barrels it is common to ignite the inside surface and let it burn for few minutes, which causes a film of black charcoal to develop on the inner surface. Such toasting is called "charring." Back to wine barrels, in many cases the heads are also toasted before being placed in position. The primary reason for toasting is to soften the phenol extraction of the new barrel by decomposing some of the gallic and ellagic tannins and the polysaccharides near the barrel surface. The heat also decomposes the wood lignin to release some volatile phenolic (see later in this chapter, and in "Specific Subjects," chapter 7, section A). The newly toasted barrel gives a very distinct aroma, described as "fresh bread," "butterscotch," and "toasted almond," to the wine being stored in it, especially to white wine. Another benefit of toasting from the mechanical point of view is that it relieves some of the stress of the newly made barrel, thus reducing the cracking potential of the weak bent staves.

2. Barrel Evaporation

It is well known that wine stored in barrels does evaporate, and continuous topping is necessary to keep the barrel full. Water and ethanol diffuse into the wood, carrying with them some nonvolatile components. A cross section of old staves from a red wine barrel shows red pigment in the inner side of the staves up to a few mm. The penetration of nonvolatile compounds is limited in depth, but the volatile water and alcohol can diffuse all the way out from the barrel to be lost in the cellar environment. The evaporation rate depends on the temperature (higher vapor pressure at higher temperature), air movements, and on the relative humidity in the cellar. The figure at the top of the next page shows the combined effect of temperature and relative humidity on the annual evaporation rate of a typical barrel in wine lost (%).

Water molecules are smaller than ethanol molecules and are therefore more able to diffuse out. But on the other hand, water vapor pressure is lower than that of ethanol. In general, under typical barrel storage conditions, the total evaporation rate is about (2 – 6) % per year, which is about (4 – 12) L per 225L barrel. If the bung is tightly inserted, the diffusion out of water and alcohol can develop a vacuum inside the barrel. The reduced pressure may develop to about a few inches of water. The existence of a vacuum in the barrel is very good proof that air is not able to diffuse into a wet full barrel, which means that oxidation of the wine through the barrel walls is not significant. The oxidation of wine stored in barrels happens

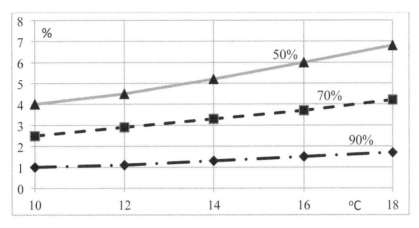

**Annual evaporation of wine (in %) from 225L barrel at
various temperatures and cellar relative humidity**

during cellar operations such as barrel racking and topping. Also, the barrel being a small volume container, any topping which is done frequently is enough to add the desired oxygen for the barrel-stored wine to age. The evaporation rate of red wine (Bordeaux varieties) stored in the same type of barrels (French barrels, Allier oak, Chateaux style, medium toast) in two storage conditions: (1) Natural conditions of a cave, and (2) An air-conditioned cellar, was studied. The temperature and relative humidity measurements showed that they were very stable in the cave (during the 50 weeks of the experimental period), while in the air-conditioned cellar the fluctuations of both parameters were drastically higher. The average temperature and relative humidity values in the cave and in the air-conditioned cellar were 16.7°C / 92.5%, and 13.5°C / 74% respectively. The mean topping wine volume (after taking a correction for expansion / contraction changes at each topping) was 2.5L / barrel per year in the cave, compared to 7.0L / barrel per year in the air-conditioned cellar. It appears that the stable conditions existing in the cave help to reduce wine evaporation. If a stable, controlled, air-conditioned cellar could be maintained, probably the evaporation rate would be the same as in a cave.

In regard to the difference in evaporation rates between water and alcohol (in the above experiment), the final concentration of ethanol in the wine kept in the cave dropped by 0.2%, while the air-conditioned cellar wine gained 0.1% (from 13% before the experiment). It appears, therefore, that the balance point, namely, equal evaporation of water and ethanol from wine kept in barrels, must be somewhere above 74% relative humidity (the mean value in the air-conditioned cellar). This finding is somewhat different from a previous report that below (60 – 65) % relative humidity, water evaporates faster than alcohol, and above that humidity, alcohol evaporates faster. In the first case (higher water evaporation), the wine became more concentrated with alcohol, and vice versa in the second case. For practical purposes, a value of 70% relative humidity can be used as the balance point where equal volumes of water and alcohol are evaporated, leaving the wine with the same alcohol concentration.

Another study confirmed the higher losses of wine under ambient conditions compared to controlled environment (temperature and humidity). The experiment was done with *topped* barrels (weekly topping) and *rolled barrels* (bung at 2 o'clock position with no topping for the entire length of the experiment). From this experiment it is evident that the losses were much higher in the uncontrolled conditions, where the temperature variations were in the range of (14 – 26) °C and storage humidity between (54 – 90) %, than in the controlled conditions, with (14 – 16) °C variations and humidity range between (85 – 95) %. Alcohol changes were like those of the previous study. At ambient conditions where humidity changed from low to 90%, evaporation rate of water was much faster than at the controlled condition with humidity in the 90% region. Also, the extra 10°C at the ambient conditions contributed additional evaporation. These two factors caused the increase from 1% to 4% loss of wine. Although there are some economic losses, this is not always bad, because wine is concentrated by a few %, with some possible quality enhancement.

3. Barrel Maintenance
Preparation of New Barrels

New barrels should be swelled properly in order not to cause any damage to the wood. Some guidelines might be helpful:

(1) Place 20 to 30 liters of hot water (60 – 70) °C in the barrel, wet the whole inside and stand vertically for two hours. Change to the other side for another two hours, rinse the water, and fill with wine. It can be done also with cold water but instead of two hours, 12 hours will be needed for each side.

(2) Occasionally you may see wet spots outside the barrel, indicating tiny leaks. In rare cases you might see more severe leakage. The barrel may leak because some staves or the heads were loosened after a long period of being dry. In such a case, fill the barrel completely and leave it overnight. In most cases the leak will stop.

(3) If not, continue to add water for a few more days until the leaking stops.

(4) In very rare cases you may see a leak at the end of the stave in the round part of the barrel holding the head. This shows the long path that the water took along a tiny slit in the wood to the end of the stave. To stop the leak, you may widen the place of the outside leak with a thin tip, and insert a sharp tip of oak wood into the hole with a hammer. Cut off the extra wood. (All this can be done while the barrel is full of water). The piece of wood you inserted will swell and most probably the leak will stop. (In a case when the barrel is full of wine and a leak develops, the same treatment can be done while wine is in the barrel.)

(5) If a new barrel still leaks after a week of soaking, contact your agent.

(6) Do not use any chemical substance except clean water in preparing a new barrel. It you use chemicals it may change the natural flavor of the newly toasted barrel.

Storage of Empty Barrels

In general it is always better to keep the barrels full with wine. When this is not possible, care should be taken to store empty used barrels:

(1) When *new barrels* must be stored for a long time before use, do nothing; simply leave them with their plastic cover as they are shipped and keep them in a cool room with relative humidity above 75%.

(2) To store *used barrels*, rinse them well with water after wine has been removed. Drain the water out and spray sulfur-dioxide gas for three to four seconds, and close the bung with a silicone stopper. Burning sulfur paper inside the barrel is another alternative for sulfiting. Cycles of rinsing (wetting the barrels) and sulfiting every month is necessary in order to preserve the barrels in good shape for later usage.

(3) A used empty barrel has to be swelled like a new barrel before wine can be placed in it. The water placed in the barrel during swelling also removes the accumulated sulfur-dioxide from the barrel interior. For this, fill the barrel with water and leave for 24 hours. Empty the barrel, rinse it, and let it drain well before filling with wine.

(4) In the rare case when the barrel was not in use for a very long time, and was not treated well, the hoops may become loose. Do not force the loosened hoops into a new position (further up the barrel curvature) to tighten the staves. If you do so, when the barrel swells, there is a good chance that some staves may crack. In most cases the barrel has finished its duty. But if you insist on saving it, the best way is to place the barrel in an open container bigger than the barrel, and fill it with water,

so that the barrel stands in water inside and outside for a few days. In some cases this works and the barrel can hold water but still leaks. Then you can take it out of the container and continue to fill it again and again until it stops leaking completely, or just leave it and do not use this barrel any more. After the barrel stops leaking, in most cases the hoops will be tight in the right position. You can then try to reposition them slightly with a hammer.

Using Chemicals in Barrel Maintenance

This is not recommended, because chemicals may leach out the barrel flavor, and may change the grain structure of the wood. They should be used only in severe cases of microbial spoilage that caused undesirable odors in the barrel:

(1) The most drastic chemicals for such treatment are caustic soda (NaOH) or soda-ash (Na_2CO_3). The concentration should be between (100 – 400) gr/barrel of 60 gallons. Do not use higher concentrations. Dissolve the caustic soda in water, and add to the barrel when it is half full of water. Then add more water to fill the barrel. After a few hours, empty the barrel and rinse with water several times. Neutralize the residual caustic soda with citric or tartaric acid with about 100 gr/barrel. Caustic soda is highly basic (pH above 13) and **care should be taken when dealing with this chemical**. Especially avoid any contact with the eyes!

(2) Chlorine treatment (or iodine) is very effective in sterilizing microbial organisms. The recommended concentration is 200 ppm of active chlorine or (15 – 25) ppm of iodine. There are several chemical sources for chlorine (and iodine) and the user should look for the producer instructions. The chemical has to be dissolved in water before adding it to the barrel. It should stay in the barrel for a day or two. Then the barrel should be rinsed with water several times, and be neutralized with citric or tartaric acid. There are some questions about using chlorine to sterilize barrels, because of the possible reactions between it and the barrel phenolic which may lead to undesirable, off-flavored compounds.

(3) Citric or tartaric acid is used to neutralize any basic residual chemical left in the treated barrel. The recommended concentration is 0.5 gr/L. It should be dissolved in water before being added to the barrel, then be left for a few hours. After removal, the barrel should be rinsed with clean water, then filled with wine or be sulfited for storage.

(4) In serious cases of bacterial infection, e.g., acetobacter, or an empty barrel after a long time of storage which has a very strong smell of acetic acid, indicating

that the barrel is infected, special treatment is needed. In recent study on the various treatments for sanitizing such barrels, the options studied were:

(a) Chlorine solution (250 mg/L of household bleaching solution);

(b) SO_2 solution (250 ppm or more and tartaric acid to pH = 3.04);

(c) Sodium-ash (1.25 g/L);

(d) Hot water at 85°C.

Only the final treatment with hot water was found to achieve the best results. The chemicals were in use for 24 hours, while the hot water was used for only 20 minutes. The reason why the chemicals were not able to kill all the bacteria cells is most probably that they were not able to come in contact with the bacteria inside the wood pores. On the other hand, the *heat* of the hot water does transfer through the staves' surface into the pores, to reach the bacteria cells and to kill them all.

It is also possible to use ozone gas inside the barrel for say an hour and it works very well. Another technique which can sterilize barrels, mainly against *Brettano-myces*, is an ultrasound wave machine which penetrates into the small pores in the wood and kills the bacteria. By doing so, it also cleans the tartaric acid deposit on the barrel's walls very effectively.

Tartrate Removal

When tartrate has been deposited on the inside of a barrel:

(1) It is best to remove it by soaking with warm water (40°C). Repeat the soaking several times until the warm water has dissolved all the tartrate. Using a high pressure spinning ball inside the barrel is also efficient for cleaning the hard layer of tartrate.

(2) If very heavy tartrate is deposited inside the barrel, caustic soda can dissolve it (see above). Ultrasound waves are also suitable for this purpose (see above).

Maintaining Full Barrels at Storage

Wine stored in barrels should be well-maintained in order to minimize any possible damage to it:

(1) Due to wine evaporation, top the wine at time intervals that do not allow the added wine (at each topping) to be more than (0.5 – 1.0) L in a 60-gallon barrel. After filling wine into a new barrel, the first topping time may be quite short (a week or two). After the first or second topping it stabilizes for the rest of the aging time.

(2) If wine leaks, but *not* so seriously that it can't be kept in the barrel (which

usually will seal itself after a while), clean the outside surface with SO_2 solution, so that no mold will grow on it. A leaking barrel can also be repaired by pushing oak sticks into the hole. If the leak is serious and does not stop, it will be necessary to replace the barrel. (According to Murphy's Law, the faulty barrel will be the lowest barrel in the stack, with several barrels above it.)

(3) Wine stored in new barrels tends to extract oak flavor very fast. Frequent checking and tasting of the wine is strongly recommended (especially with white wine) in order not to over-oak the wine.

(4) Red wine stored in barrels should be held between six months and two years, depending on the wine quality, barrel age, and storage conditions. When white wine is aged in barrels, the length of time should be much shorter, from a few weeks up to several months. Storage time depends very much on the storage temperature. At higher temperatures the extraction rate is higher, so the storage time should be shorter.

Life Cycle of Barrels

The length of time before the barrel exhausts its flavor is about four to six years. The length of time needed to age wine in a barrel, to extract the desired flavor, gets longer as the barrel becomes older. In many French chateaux, each vintage is held in new barrels for a period of (18 – 24) months. Afterward, the barrels are sold.

(1) A good practice for starting the life cycle of a new barrel is to ferment white wine in it. The yeast softens the barrel tannin for the next refill. Also, the new toasted barrel gives the white wine a roasted almond flavor, which is very compatible, for example, with Chardonnay or Sauvignon Blanc wines. In some cases the wine is left on the yeast (*sur-lie*) for an extended period of a few months, and is then blended with stainless-steel-fermented wine. In certain wineries in Bordeaux, Sauvignon Blanc is left over *sur-lie* in new barrels for one to two years, with occasional stirring of the lees. The yeast lees protect the wine from becoming over-oaked. Instead, the wine becomes very round and full-bodied, with a surprising fresh-fruity aroma and extremely complicated bouquet.

(2) Controlled temperature and humidity are most recommended for barrel storage.

B. Barrel Aging

The use of oak barrels to age red and some white wine is well established. The overall effects on red wine storage in barrels are:

(1) Slow and controlled oxidation which softens the wine's tannin and increases the red color intensity and stability by condensation between anthocyanins and other phenolic compounds.

(2) Extraction of oak phenols and flavor components from the barrel which enhances and expands the wine's complexity.

(3) Evaporation of water and alcohol at an annual rate of (2 – 6) %, which in fact increases the wine's dry extract concentration and its flavor.

(4) The gradual development of an aged wine bouquet.

1. Oak Components and Extraction

Some basic information on the composition of oak extraction components is important for understanding the wine/barrel relationship.

Phenolic extraction

Tannins: The most extractable phenolic from oak barrel are the nonflavonoid phenols such as *gallic-acid*, *ellagic-acid*, *lignin*, along with their degradation products. The total phenols extraction was measured in new French and American barrels filled with white wine (Sauvignon Blanc). The wine was aged in the barrels for about three months. Then the barrels were refilled with new wine for the same period. The quantities of phenolic extractions in the two series of barrel aging are shown in the following figures:

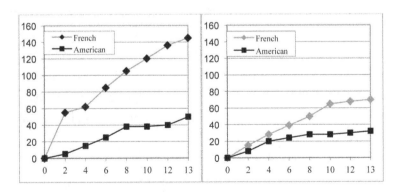

Barrels' first fill   *Barrels' second fill*

Phenolic extraction from French and American barrels during first and second filling with white wine (in mg/L). Time scale in weeks.

It is very clear that the French barrels contain more extractable phenols than the American barrels. Notice also the difference in the extract concentration between the new and used barrels. In the second refill the French barrel extract dropped considerably, while in the American one, it is more or less the same. This pattern has been shown in other studies as well.

The phenolic compounds extracted from barrels can be classified into *nonvolatile phenolic,* which might contribute *astringency* to the wine, and *volatile phenolic,* which enriches the wine with a distinctive *oak bouquet.*

***Nonvolatile Phenolic*:** The astringency *thresholds* of oak phenolic in wine were carefully studied. It was found that the astringency *thresholds* of seasoned oak extract containing *only* nonvolatile phenolic (such as gallic and ellagic acids) are: (800 - 1600) mg/L (as Gallic Acid Equivalent) when extracted from French oak, and (200 – 400) mg/L from American oak. This study focuses on the difference in the astringent character between French and American barrels. These findings were compared with the actual concentrations of phenolic extracted by white wines and model solutions aged in wood for 12 months. The *total* phenolic that was extracted during that period was in the (200 – 300) mg/L range from French oak, and (100 – 200) mg/L from American oak. The conclusion is obvious, that there is practically no significant addition of astringent phenolic to wine from barrel aging. Or in other words, barrel aging *does not add an extra astringency* to wine. These findings were also supported before. (The reader should remember that these are nonvolatile phenolic which do not contribute to the nose perception.)

***Volatile Oak Phenolic*:** Known also as *Aromatic aldehydes*, they are the main degradation products of lignin. The degradation may take place either in the acidic wine medium, or thermally when the barrels are toasted. The major compound is vanillaldehyde (vanillin), which gives to the wine a typical vanilla flavor. Regular concentration levels of vanillin in oak aged wines are much above its taste threshold (which is about 0.5 mg/L in 10% ethanol solution). Syringaldehyde, coniferylalde-hyde, and sinapaldehyde are some other extracted aromatic aldehydes:

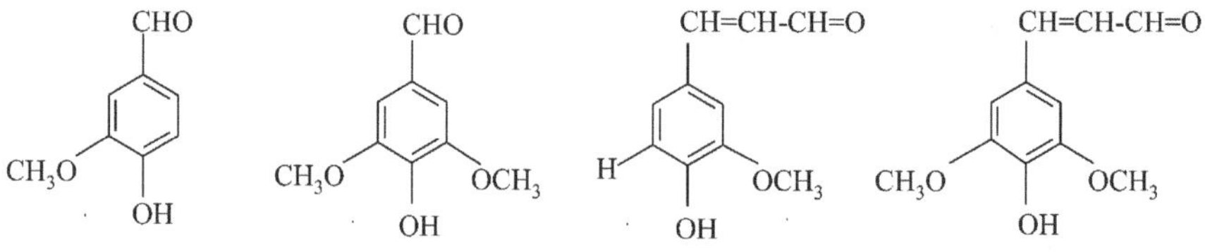

vanillin   syringaldehyde   coniferaldehyde   sinapaldehyde

The toasting level will determine to what extent such aromatic aldehydes will be potentially extractable from a new-made barrel.

Aromatic aldehydes are formed also by lignin degradation in the ethanol-acidic solution of wine. A very profound effect of lignin degradation was shown in old barrels containing Armagnac for twenty years. The inner and outer faces of the staves were cut (2 mm thickness) and the aromatic aldehydes content in the wood was measured. The results are shown in the following figure based on the above data:

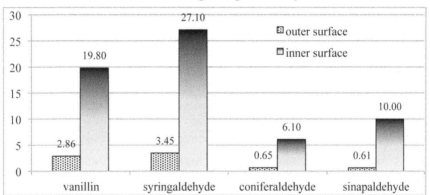

Aromatic aldehydes content in the outer and inner surfaces of oak barrels containing armagnac for 20 years

The data show that lignin in contact with the alcohol-acidic solution is decomposed to its basic aldehyde components, which diffuse out from the wood into the liquid. Notice also the very high concentration of syringaldehyde.

The concentrations of aromatic aldehydes during barrel aging were also measured with Cabernet Sauvignon wine, in two barrel sizes, 225 and 500 liters. The results are shown in the following figure based on the above data:

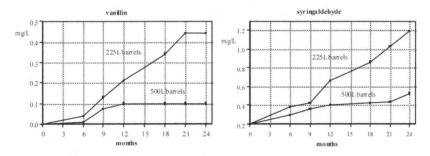

Vanillin and syringaldehyde concentrations in Cabernet Sauvignon wine aged in oak barrels of 225L and 500L volume size. The same wine aged in concrete tanks showed only traces of this aldehydes.

The effect of the smaller barrel size on the extraction rate is very clear.

Other identified aromatic *phenolic components* were found, such as:

(1) *Eugenol,* which has a spicy clove-like flavor (and in fact is the main clove flavoring component), is extracted mainly from untoasted oak. Its highest concentrations are found when the wood is cut just before seasoning. In seasoned wood its typical concentration is about (2 – 3) mg/Kg (a reduction from a typical value of 6 mg/Kg in green timber). The odor threshold of eugenol in white and red wines is 0.18 and 0.70 mg/L respectively.

(2) *Guaiacol* and *4-methyl* and *4-ethyl guaiacol* are degradation products of thermal decomposition of lignin, and have a smoky flavor.

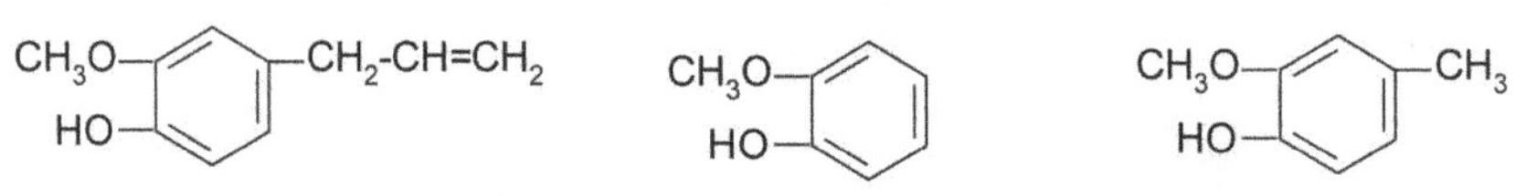

| eugenol | guaiacol | 4-methyl-guaiacol |

The odor threshold of guaiacol in white and red wines is 20 mg/L.

(3) *γ-lactones*, compounds with very significant contribution to oak flavor, are two isomers *cis* and *trans-β-methyl-γ-lactones* (4-methyl-5-butyl-dihydro-2-furanone), or simply called **γ-lactones**. They were identified first in oak extract and later recognized as important flavor compounds in whisky, brandy, and wines aged in oak barrels. The *γ-lactones* are dia-stereoisomers (that is, they have two asymmetric carbons in the molecule), and hence there are four isomeric molecules, two in each of the *cis* and *trans* forms:

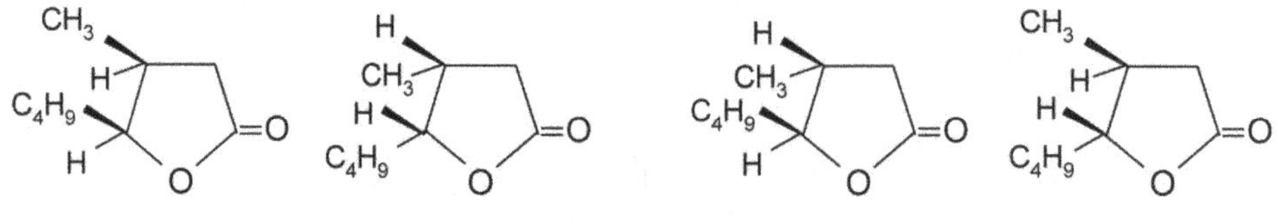

cis-β-methyl-γ-lactone   trans-β-methyl-γ-lactone

After identification of these lactones, there was some confusion in the literature regarding the correct isomeric assignment. For example, odor thresholds and other physical characteristics of the two forms cis and trans were given to the wrong isomer. The absolute configurations (shown here) were confirmed later.

The odor thresholds of the two γ-lactones are quite different. The detection threshold of the *cis* isomers was found to be 0.067 mg/L, and of the *trans* isomers is

0.79 mg/L (in 30% alcohol solution). The woody, coconut, oaky aroma of lactones is therefore attributed mainly to the *cis γ-lactones* (tenfold lower odor threshold); γ-lactone concentration in oak wood is in the (10 – 50) mg/kg range. It is worth mentioning that there are also other lactone compounds which are found in wine, and which are not contributed from oak aging (mainly in sherry).

White Wine Sur Lie: The effect of fermenting white wine in barrels and leaving it on its lees for few months (over a year in certain cases) was compared with another alternative, such as tank fermentation and then racking and yeast removal before barrel aging for eleven months. In most extraction aspects, there was no significant difference between the two cases in the composition of most compounds under study, including *eugenol, guaiacol, 4-methyl-guaiacol* and *γ-lactones*. However, there was one significant difference. It was the vanillin content, which was much lower in the *sur lie* wine. This was caused by the yeast biomass, which interacts with vanillin, leaving a very small amount of it in the wine. The interaction takes place during fermentation and afterwards. This interaction explains why barrel fermentation (and *sur lie* aging) does not impact much of the oak character of the wine (although in most cases, new barrels are used). This effect was previously shown, where with added oak extract to fermenting model wine, the vanillin practically vanished. Or in practical winemaking, wines fermented and aged in barrels had lower content of vanillin (and vanillin flavor) than wines fermented in stainless steel tanks and then aged in barrels for the same time.

2. French vs. American Oak

The question of French vs. American barrels has a significant economic impact on wine production. Assuming three years of barrel usage (a 60-gallon barrel which is equivalent to 300 bottles) and an average storage time of one and a half years for each vintage (namely, 600 bottles per barrel lifetime during two vintages), the difference in oak cost of French over American barrels is about $.07 per bottle. It is not very significant in high-priced wine, but it is significant in the fighting varietal segment of the market. So what are the real actual differences between the two barrel categories?

There is a basic botanical difference in the oak species as mentioned above. The French barrels are made of *Quercus-robur* and *Quercus-sessilis*, while most of the American barrels are made of the *Quercus-alba* and other American species such as *Q.-bicolor, Q.-macrocarpa, Q.-prinus, Q.-stellata*, and *Q.-lyrata*. Other factors such as soil and climatic differences, growth rate, cutting location of the staves in the tree, and cooperage practice (wood drying, stave-cutting methods, fire or steam bending, and toasting methods and level) also contribute to the differences between the two categories of cooperage. A basic difference was observed in the width of the annual rings between the main oak species. The American oak annual rings are in general wider and more variable in width [(1 – 5) mm/year] than the French *Sessile* ones [(1 – 3) mm/year], but thinner than the French *Robur* oak [(3 – 10) mm/year]. In each year two rings are produced; one is lighter in color and wider (spring), and the second is darker and thinner (winter). It was also found that American oak can be *sawn* along the logs to make the staves, instead of splitting them, with no leakage problems. The wood structure is denser, so there are fewer leak channels through the radial lines. This is of great importance because more wood per tree can be used to make barrels. One cubic meter of wood logs can produce two barrels by splitting, whereas by sawing, four barrels can be produced. This is one major parameter for the price difference between French and American oak barrels.

Studies on oak extract, either as total solids or total phenols, show a remarkable difference between the two. French oak (or in general European oak) contributes more solid extracts and phenols than American oak. The results in one of the studies, which measured the total phenol compounds and the non-flavonoid phenols, extracted from 1 gram of wood (as chips) per 1L of solution, are shown in the following figure:

Total and non-flavonoid phenolic extraction from 1 gram of
French and American oak chips in 55% ethanolic solution

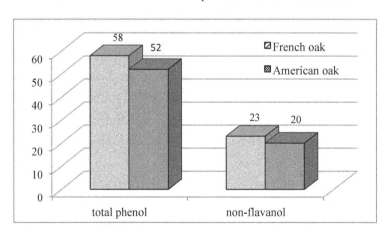

(The extraction was done with 55% alcoholic solution,
which represents brandy aging in wood, but not wine.)

The difference is quite high. About half of the French oak chips (compared with American oak) were then considered to be necessary in order to give a treated wine a detectable threshold of oakiness (1.1 gr/L of French oak chips vs. 2.2 gr/L of American oak ones). This is not really the actual case, because when specific non-flavonoid analysis was carried out (in a different study), although gallic-acid, protocatechuic-acid, synaptic-acid, caffeic-acid, and total phenols were found to be higher in the French oak species (*Q.-robur*), the concentration of *vanillin* was found to be *higher* in American oak (*Q.-alba*). But the last compound has the highest impact on oakiness compared to other oak components, and therefore American oak should actually give wine a more "oaky" impression than French oak. On the other hand, as the total phenols are higher in French oak, the oak "*astringency*" is expected to be more noticeable in the French barrels (which is not the case because the extracted nonvolatile phenolic are below their taste threshold; see above in this section).

Specific measurements of the major *non-flavonoid tannin* showed also that a higher extract was found in French oak, in comparison with American oak. Details on some other differences in basic components of wine stored in the two different oak categories were studied in another work. Chemical analysis of Cabernet Sauvignon wine aged in French and American barrels (same cooperage method such as air-dried, fire bending and light toasting) revealed the following results:

(1) Phenolic extraction (total) from French barrels was measured to be only slightly higher than from American ones, as can be seen in the following figure based on data from the above study:

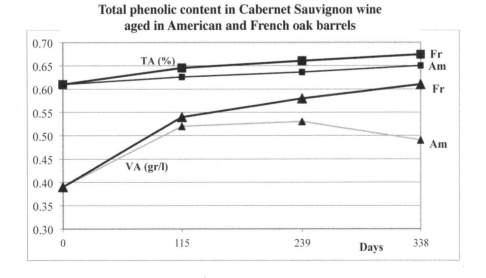

Total phenolic content in Cabernet Sauvignon wine aged in American and French oak barrels

In this case, the accumulated phenolic extraction during almost one year from the American and French barrels was 2% and 5.5% of the total wine phenols, respectively.

(**2**) Titratable acidity (TA) increased slightly more during aging when the wine was aged in French oak compared to aging in American oak. The rise of the TA concentration is attributed to evaporation from the barrels, and to acid extraction from the wood. The pH has a decreasing tendency over the same period (in both barrels), but it is not significant.

(**3**) Volatile acidity also increases over aging, with significantly higher concentrations in the French barrels. The increase in VA is related to volatile acids extraction from the wood. These results can be seen in the following figure based on the above study:

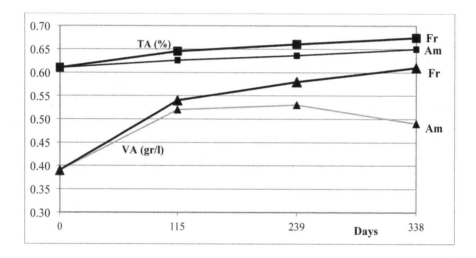

There is no comment on potassium bitartrate-precipitation during the aging period, which may have some influence on the TA and the pH changes.

(**4**) Of greatest interest is the finding (in another study) that in both kinds of barrels, there was not any significant difference in wines that were aged for 115 days (4 months) in comparison to 338 days (11 months). Sensory descriptive analysis in model wines treated with oak chips of various origins revealed that American oak is perceived to be less intense in its spicy, nutty aroma than the French oaks. Many studies have been done on various aspects of this subject. The results are still controversial and unclear, and more has to be done to clarify this issue.

3. Oak Alternatives

To shorten the time and expense of wine aging, experiments were done with oak chips or oak dust immersed in the wine. This "fast-aging" method is not illegal, and if it is suitable to a specific wine, there is no reason why it should not be used. Even simulation of wine aging and brandy aging, by alcoholic (or water) oak extract, has been suggested and patented.

This practice of using oak alternatives like oak chips, oak shavings, oak staves, or any structure of oak used in the wine industry, became legal in the European Community in October 2006. The replacement of oak barrels by oak chips can fulfill only the extraction function of barrel aging, and not the oxidative one. To achieve the other function of barrel aging, controlled aeration has to take place. Here we will focus our attention only on the extraction segment of the treatment. (For more details on the whole subject, which is called *micro-oxygenation*, see chapter VII section B.)

The size of oak pieces used in the industry can be roughly divided into: *powder* (less than one mm particles), *shavings* (thin particles of about 1 mm thick and a few mm length and width), *and cubes* (about 10 mm and above), and *random shapes* (big shavings of a few cm in size). Besides the small particles, staves of various sizes and (20 – 30) mm thick are placed in stainless steel tanks or used barrels. In this case they act very much like barrel staves. The amount of American oak chips needed to be immersed in red wine in order to produce the oak *recognition threshold* was estimated as 2.5 g/L of wine; for a full barrel aging of red wine it would require an American oak chips treatment of about 15 grams of oak chips /L of wine. The common quantity of small oak pieces used in the wine industry is on the order of (1.5 – 2.5) g/L of wine, for a period of a few weeks up to a few months. In one study it was shown that with small oak pieces, one week is practically enough to get extraction equilibrium.

Recent papers on this matter, designed to collect more accurate data on how to use oak alternatives, were more specific in regard to the major volatile compounds that are extracted from the wood into the wine. The compounds studied are (see earlier in this section): *cis γ-lactone, vanillin, eugenol, guaiacol* plus its derivatives, and some others. All these compounds contribute the oaky flavor to wines. Parameters such as size of the oak chips, toasting level, and time length of extraction were determined in order to get the best results. One of the studies measured all the above components, comparing new American barrels, used American barrels, and used American barrels with American oak chips of various sizes. One component, *cis γ-lactone*, is known to be the most recognized component with oaky flavor (coconut). This compound is also chemically stable in the wine medium, which makes it a good marker to study the extractability from the various oak options. This is shown in the following figure based on the above study:

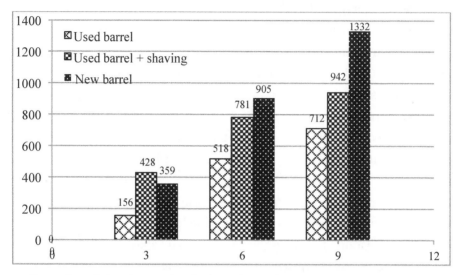

Extraction of cis γ-lactone (mg/L) by red wine from three sets of oak sources: new barrels, used barrels, and used barrels with oak shavings (3x8x1mm, 2.6 g/L) in three, six, and nine months periods. All sets are American oak.

The contribution of the oak shavings is seen clearly as the *difference* between the used barrel experiments (without and with oak shavings), which remains the same after the first 3 months. This means that equilibrium was set before that time period, and it is not necessary to leave the shavings any longer. On the other hand, with barrels the story is different. The new barrels continue to export γ-lactone to the wine in an almost linear line for the full period of the study at an average monthly rate of 150 mg/L. The used barrels (7 years old) are still linearly extractable, with about half the rate of the new barrels. This is because it takes longer for the wine to diffuse into the depth of the wood and for the wood components to migrate into the wine. The absolute numbers of extraction are less important than the changing of the diffusion rates from the barrels, with their age. The same representation could be done on another component, e.g., vanillin, but it will not serve the purpose because vanillin is an aldehyde and is oxidized to its alcohol and other derivatives, so it is not a stable marker like γ-lactone. The extraction of the same volatile compounds from oak pieces was demonstrated in another study in model wine solutions and in wines. In this study, 10 g/L of oak chips (4 times greater than the previous study) were used, and all components' concentrations measured were very much higher. The maximum extraction times were reached in four months, with γ-lactone concentration at about 5 times greater than the previous oak extraction study with 2.5g/L of oak pieces.

A sensory comparison was made between white wine aged in three different conditions, namely, new American barrels, used American barrels plus added oak chips, and stainless steel tanks plus oak chips. All the wines were aged to contain the same extracted quantity of gallic-acid (as a probe for equal non-flavonoid extraction level). The sensory tests were very interesting. The aroma and flavor ratings of these different wine treatments were practically equal. Also, the preferred quality rating was the same for all three wines which raises some questions. *Additional and very careful research is needed to study the question of traditional barrel aging versus stainless steel plus oak chips.*

In practice, oak chips are sold in granular form as French and American oak. The chips are also available as light, medium, and heavy toast chips.

Another alternative to the traditional oak barrel is innerstaves which are installed in stainless steel tanks. This option is found in various shapes and tank sizes, but in principle the innerstaves replace the barrel as a container and function only as oak wood material for slow extraction. Also on the market are 225L stainless steel drums containing interior oak staves held by a circular stainless steel holder. The staves can be replaced after one or two years of use. Another option is to replace the drum heads with oak staves, which increases the wood area by 25%.

Shaving old barrels has also been done for many years. Usually 5-year-old (or older) barrels can be shaved and reused for another two years with good results. The main problem that may arise after shaving is that the barrel's mechanical strength is weakened, so that some staves may crack and leak, mainly at the bottom.

CHAPTER VI

BOTTLING

CHAPTER VI: BOTTLING

A bottle of wine is the final product of the whole process of winemaking. The prospective customer, before buying a wine, will have an eye on the bottle, and part of his/her decision to buy a certain wine will depend on how it looks. Furthermore, when the bottle is opened for tasting, it must meet most of the customer's expectations. In addition, the wine is kept in the winery cellar or warehouse for a certain period of time, and then at the store and at the customer's home before it is opened and consumed. Care must be taken to ensure that the wine will stay alive, and that it will age in the bottle without spoiling. All these factors must be taken into account when planning for the final operation in the winery—bottling. This chapter covers some important aspects of bottling principles which are common at any bottling scale, from tiny amateur winemaking up to full-scale winery operations. The only difference is in the equipment (hand-operated, semi-automatic, or fully automatic machinery), which is determined by the quantity of wine to be bottled.

A. Wine Container

1. Bottles

Traditionally, wine bottles are made in three major shapes: the *Bordeaux* type, the *Burgundy* type, and the *Alsace* (also *"hock"*) type.

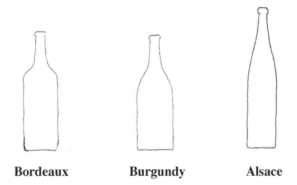

Bordeaux   **Burgundy**   **Alsace**

These names, of course, indicate the origin of these shapes. According to the tradition, each varietal wine is held in a certain type of bottle. This means that most of the world's wines are bottled in one of the above three shapes, according to the regional tradition of the major French appellations. It would seem odd, for example, to find Cabernet Sauvignon in a Burgundy bottle or Chardonnay in Alsace or Bordeaux bottles. However, one does find some exceptions to this general acceptance. For example, a bottle of Cabernet Franc from the Loire valley is bottled in Burgundy-shape bottles, in accord with this regional tradition for red wine.

The colors of the Bordeaux and Burgundy bottles are green or brown for red wines, and white or greenish-white for white wines. The Alsace bottle, which is suitable only for white wines, comes in two colors: green in the Mosel style, and brown in the Rhine style.

It is important to mention that light may cause some photochemical reactions in wines, which will accelerate wine aging. Red wines which contain high concentrations of anthocyanin and other phenolic compounds are more susceptible to these reactions. Therefore, wine that is to be aged for a longer time should be kept in a colored bottle. This is the reason why red wines are traditionally bottled in deep green or brown (Italian style) bottles. Some white wines, specifically those that are intended to age for longer time in the bottle, may also be bottled in green bottles (Chardonnay, Riesling, and others). The types of bottles, and some representative varieties, are summarized in the following table:

Bottle type	Varietal wine	Bottle color
Bordeaux	Cabernet Sauvignon	green
	Cabernet Franc	green
	Merlot	green
	Nebbiolo	green/brown
	Sangiovese	green/brown
	Zinfandel	green
	Barbera	green/brown
	Sauvignon Blanc	white
	Semillon	white
	Muscadet	white
	Muscat	white
Burgundy	Pinot Noir	green
	Sirah/Shiraz	green
	Gammay	green
	Grenache	green
	Chardonnay	white/green
	Viognier	white/green
Alsace	Riesling	green/brown
	Gewürztraminer	green/brown
	Sylvaner	green/brown
	Muscat	green/brown

Varieties that are closely related to those cited in the table are traditionally bottled in the same type of bottle as the variety they are directly related to. For example, wines in the Riesling family, such as Emerald-Riesling (Riesling x Muscadelle) or Müller-Thurgau (Riesling x Madeleine-Royale) are bottled in Alsace bottles. Practically all white German wines are bottled in Alsace bottles, because most German white varieties are Riesling-related. Cabernet Franc, Malbec and Merlot, which are associated with the Bordeaux region, are bottled in Bordeaux bottles. The same rational holds for all Pinot related varieties (Pinot Noir, Pinot Gris, Pinot Blanc, Pinotage) which are bottled in Burgundy bottles.

As for generic wines, there is no agreement which bottle is suitable, but in many cases, the regional tradition dictates the bottle style. For example, Chianti wines are bottled in Bordeaux style (although there is a unique "Fiasco" shape bottle for the simple Tuscany wines.)

The common wine-bottle size by worldwide agreement is 750 milliliters. There are also half bottles (375 milliliter), magnums (1.5 L), double magnums (3.0 L), and Imperials (6.0 L.)

The neck of all bottle types and sizes (except the Imperial) has standard dimensions in order to fit the standard cork size. At the opening side of the bottle, the internal diameter is (18 – 19) mm, and at 50 mm down inside the bottle, it is (20 – 21) mm (see figure in the next section on cork.) So the inside of the bottle's neck is tapered upward to the top. This configuration of the neck holds the cork firmly in place.

2. Corks

In principle, for fresh fruity wines that are to be consumed young, there are cheaper alternatives, such as plastic corks, crown caps, or aluminum screw caps. However, in the case of high-quality wines, sold at medium to high prices, one can afford to use the more expensive traditional cork. Pulling the cork from a bottle of wine is part of the ritual of wine serving and drinking. It is as much a part of the style and appearance of the product as the right bottle, its color, and the label. The winery's logo is also usually printed on the side of the cork.

The wide use of cork as a wine stopper started only at the middle of the 17th century. It is related (truly or not) to a monk named Dom Perignon, who looked for a solution to secure his sparkling wine in a closed bottle. Since then, the innovation of cork has become one of the most important contributions to the development of the wine industry. For many years of bottle aging, no material is known to be better than cork. However, after a long time, even the cork may deteriorate and fragment, causing the wine to leak out and allowing air to enter. Very expensive wines that are aged by collectors for many years must be opened and re-corked after (25 – 30) years.

General Aspects

The English word *cork* probably comes from the Latin word *cortex* which means "bark." The cork is made of the bark of the oak species *Quercus-suber* and *Quercus-occidentalis*, which grow mainly around the Mediterranean Sea (Portugal, Spain, France, Italy, Morocco, Algeria, and Tunisia).

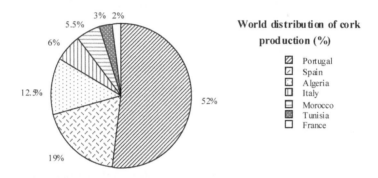

World distribution of cork production (%)

- Portugal
- Spain
- Algeria
- Italy
- Morocco
- Tunisia
- France

A new tree has to grow for about (20 – 25) years before the first bark can be peeled. It is then very dense and its structure is irregular, and therefore of poor quality and not suitable for use as a wine cork. The second bark is peeled off after (8 – 12) years. It is also of poor quality and therefore of little use. Only the bark of the third harvest is good enough to be used as a quality wine cork. That means that about (40 – 50) years are needed for a new oak tree to produce good quality corks. Nowadays, very few firms would start such a business. The time interval between each bark harvest is approximately (9 – 11) years, which is needed for developing suitable bark thickness. The number of annual rings can be seen on the flat cut of the cork. The useful life length of an oak tree as a cork source (from the time of the third bark removal) is about (100 – 150) years, although the tree may live for many centuries.

Cork Production

The production of cork stoppers is a long process. The bark is peeled from the tree and allowed to dry and cure in the open air for several (six or more) months.

It is then boiled for one to two hours to soften its texture, and to relieve its internal stress, which changes the original bent pieces to flat bark bars. The boiling eliminates insects and other living objects from the bark, and also removes volatile phenols and other volatile substances.

After a rest period of a few weeks, the bars are cut into strips whose width is the desired cork length. The strips are punched with a drill of the desired diameter to obtain the cylindrical corks. The raw corks are washed, bleached and vacuum-dried to suitable humidity (5 – 8) %. The corks are then graded by mechanical and/or manual screening to several cork grades based on faults and defects on its surface. After grading, the corks undergo surface treatment with various food-grade coating materials (paraffin, silicon oil, wax). The object of surface coating is to ease the insertion of the cork into the bottle, and also to ease its removal when the bottle is opened. The corks are then sealed in plastic bags containing SO_2 gas to ensure some sterility on shipment. As mentioned above, the corks are graded on a scale from poor quality to very high. Most of the cork producers grade their corks into (6 – 8) quality categories. The average percentage distribution of cork grades is shown in the following figure:

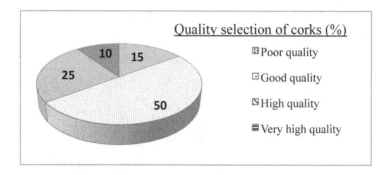

Physical properties of cork

The geometry of cork cells and the air content inside, about (60 – 85) % of its cellular volume, reflect most of its mechanical and physical properties, such as low density, compressibility, impermeability, elasticity, adherence, and heat insulation.

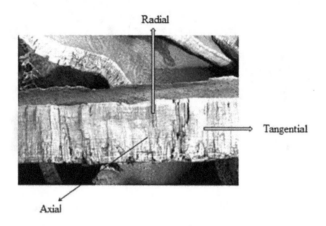

A cut through the three axes of the cork, namely radial, axial, and tangential (see figures), and a microscopic study, shows its cells structure.

Electron microscope photograph of radial (left), axial (center), and tangential (right) sections of cork.

The radial face shows a hexagonal structure like a honeycomb, while the axial and the tangential faces are seen like a stack of bricks in a wall (see the microscopic figure). The radial face is not always perfect, and other polygons cross-sections are also seen. Typical dimensions of the hexagonal prism cells are:

Wall thickness − 1 μ

Prism edge length − 20 μ

Prism length − 40 μ

Cell volume − 5×10^4 μ^3

Number of cells per 1 cubic centimeter: approximately (10 − 40) million cells.

When making wine corks, one would expect the cork to be cut through the radial face, to maintain the symmetry along the cork and yield a better seal. Unfortunately, this is not the case because of the *lenticels*, which are tiny channels connecting the outer and the inner sides of the bark. If the cork is cut in the radial direction, the wine will probably leak out. Also, if the bark is cut in the radial direction, there will be not enough width of bark for the common (35 − 50) mm cork length. Therefore, the corks are cut from the bark in the axial/tangential direction.

Summary of cork's properties

(1) **Low density** − The density of cork is in the range of (0.12 − 0.20) g/cc. The density of good quality cork is in the range of **(0.16 +/- 0.02) g/cc** and therefore, the weight of high-grade corks of 24 mm diameter (standard dimension) by 44 mm (regular length) and 49 mm (higher length) are about 3.2 gram and 3.6 gram respectively. Densities higher than 0.18 g/cc indicate low elasticity and poorer quality.

(2) Compressibility: The ability to be compressed without permanent change of the original structure. A typical stress-strain curve of cork material is shown in the figure. There is some linear strain at low stress (up to point **a**.)

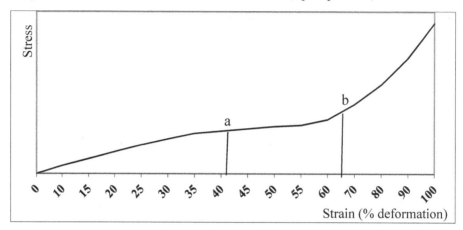

At a certain point on the stress/strain curve, where the cork's cell walls collapse because of high stress, there is higher deformation at a lower extra stress (between points **a** and **b**.) At very high deformation the stress/strain curve rises very sharply (above point **b**.) In this region, very high stress is needed to make a deformation (strain). On the linear section of the curve (between zero to **a**), the deformation of the cork dimensions is still reversible. This is one of the most important characteristics of good quality cork. The cork should recover about 95% of its original shape in a very short time (a few seconds) after being released from compression. In the actual process of wine corking, the cork is squeezed and pushed into the bottle, and because of the smaller diameter of the bottle's neck, the cork is not able to regain its original size. Therefore, the internal pressure (within the cells) is not released, and thus the cork adheres firmly to the bottle's walls.

(3) Elasticity or Resilience: A fast recovery from deformation made by compression. Under stress, the air inside the cells is held compressed, and when the stress is removed, the pressurized air in the cells exerts force to restore the original shape. In cases where the cork is being stressed for a long period (as is the case when the cork is in the bottle), partial air content inside the cells diffuses slowly out, and when the stress is removed, the cork does not entirely regain its original shape and size. A bottle's neck diameter is about 18 mm, and the diameter of most corks is 24 mm, which means a reduction of the bottled cork radial dimension by about 25%, or 45% of its volume (which is between points **a** and **b** in the stress-strain curve above). When the cork is pulled out after being in the bottle for a long period, it does not regain its original size. In sparkling wine corks, where the original cork diameter was about 30 mm, the volume reduction in the bottle is almost 65%, and when it is pulled out it gets the typical mushroom shape of sparkling wine corks.

(4) Impermeability to liquids: Resistance to liquid penetration through the cork structure is most important.

(5) Permeability to gases: Slow penetration of gases through the cork, by presumably microscopic channels ($\sim 0.05\mu$) in the cell walls. At the diameter of such channels liquid cannot penetrate, but gases and vapors can slowly pass through. It was claimed that when the cork is compressed in the bottle, the condensed air inside the cells acts as a barrier to gas penetration. On the other hand, it is well known that after a certain time in contact with wine, corks gain weight by absorbing water vapors. Measurements on absorption of other vapors (phenol and chloroform) showed a remarkable increase in weight (about 30%). It appears that the subject of the permeability of gases through corks needs further investigation.

(6) Adherence: After being compressed and pushed into the bottleneck, cork has the property of strongly adhering to the glass, because of the desire to expand, caused by the squeezed cells.

(7) Insulation: Cork has excellent heat insulation, due to the air held inside the cork structure.

(8) Immunity: Cork is very resistant to deterioration and rot decay.

All these properties make cork an excellent stopper for wine.

Cork Taint

The phenomenon of cork taint, or "corkiness" as it is sometimes called, is characterized by a moldy/musty/smoky off-flavor. Cork-tainted wines can be distinguished from wines that have off-flavor caused by microbial fault, faulty barrels, or moldy grapes. The distinction is based upon the circumstances: when the faulty wine was discovered as an individual bottle in a good sound batch, or, when the whole bottling batch is musty. The first case of *individual* musty bottles might be caused by faulty corks and is called "cork-taint." If the source of the fault is in the vinification process, then the *whole batch* will be off-flavored; the fault is not randomly distributed. Estimation of the occurrence of cork-tainted wines during the last twenty years is about 2% of all wine bottles, which seems to be a serious problem.

The components found in corks which might be responsible for cork taint are:

2,4,6-Trichloroanisole (TCA) is a major component of wine fault caused by cork.

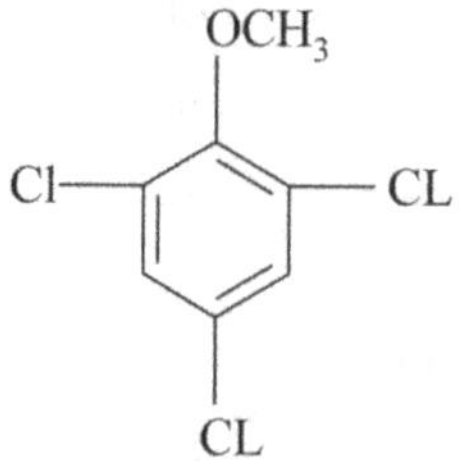

OCH₃

Cl——————CL

CL

2,4,6 – trichloroanisole (TCA)

It was recognized as one of the major cork taints in 1982, although it was known earlier as a food off-flavor component caused by fungal activity. Its musty/moldy off-flavor threshold was estimated to be extremely low. TCA was also found to be responsible for the off-flavor in coffee known as "Rio flavor." This compound was first found in cork-tainted wines at a concentration range of (20 – 370) ng/L, in comparison with less than 10 ng/L in good untainted wines. An organoleptic test made by addition of TCA to a good wine at a concentration range from zero to 100 ng/L revealed that the musty off-flavor is recognized at (10 – 30) ng/L.

TCA is formed in the cork in two steps. At first, some of the phenol components in the cork are chlorinated during the bleaching process of the corks with calcium hypochlorite to 2,4,6-trichloro-phenol, which is then methoxylated to 2,4,6-trichloro-anisole by fungal activity. To avoid such risk, most producers now perform the bleaching process with hydrogen-peroxide rather than with hypo-chloride. Another possible synthesis of TCA that was mentioned in the literature takes place in the forest by direct microbial synthesis in the wood bark from carbohydrates and a chlorine source. TCA and other taint components (listed below) were found in wines not being corked at all, but which were stored in oak barrels. The source for these contaminations was attributed to microbial activity in the barrel's wood.

o-hydroxyanisole (guaiacol) causes wine taint which is described as a burnt, smoky, and medicinal off-flavor.

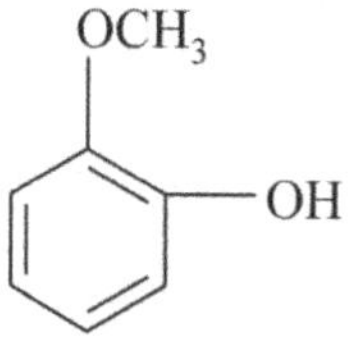

OCH₃

——OH

2-methoxyphenol (guaiacol)

Its off-flavor threshold in wine was found to be 20 µg/L, and in water 13 µg/L. In a study to find the cause of tainted wine, it was found that all wines described as tainted (9 samples) contained (70 – 2600) µg/L of guaiacol. In reference wines (3 samples of untainted wines of the same bottling) the concentrations in the wines were (3 – 6) µg/L. The traces of guaiacol found in sound wines looks to be reasonable, as this phenol was identified in trace quantities in corks, as a degradation product of lignin. It is also found in smoked meat and fish, roasted coffee, and kiln-dried malt as a phenolic pyrolysis product of wooden material (and also in toasted barrels) from lignin via ferulic-acid to guaiacol.

1-octen-3-one was also found in cork-tainted wines.

$$CH_3-CH_2-CH_2-CH_2-CH_2-CO-CH=CH_2$$
1-octen-3-one

The smell is described as mushroom-like, with odor threshold of 20 ng/L similar to that of TCA.

Trans-1,10-dimethyl-trans-9-decalol (geosmin) and **2-methylisoborneol**

Trans-1,10-dimethyl-
trans-9-decalol (geosmin)   **2-methylisoborneol**

are two other off-flavor components found in cork-tainted wines, with an earthy odor and very low threshold of 25 and 30 ng/L, respectively. These compounds are natural byproducts of soil bacteria, which are also found in water supplies. The specified isomers of these compounds (as written here) are those that are responsible for their odor. Their occurrence in cork is probably due to contamination of the cork bark when it is cured in the open air, or later in storage.

To test a new batch of corks for the presence of cork-taint components, a simple method has been suggested: randomly choose 20 – 30 corks and place in a closed beaker with one liter of neutral white wine which has not been in a wooden container. The corks should be fully submerged. After about (6 – 8) hours, smell the tested wine (with its reference wine) and look for moldy, musty flavor. To cover the whole batch, the test should be done with a set of multiple samples.

A summary of cork-taint compounds is given in the following table:

Compound	Off-flavor description	Threshold in wine
TCA	musty, moldy, wet cardboard	4 –10 ng/L
Guaiacol	smoky, medicinal, phenolic	20 μg/L
1-octen-3-one	mushroom-like	20 ng/L
Geosmin	earthy	25 ng/L
2-methyl-isoborneol	earthy	30 ng/L

It thus appears that many compounds can contribute to cork taint, and although this problem has definitely diminished, it still exists and no absolute solution is yet known. In part, this opened the door to look for other alternatives.

Practical aspects

The cork must fit into the wine bottle, according to its size and shape.
The standard diameter of wine corks is (24 – 25) mm. The common lengths of wine corks are: 38 mm for simple and short term wines, 44 mm for medium aging wines (the most common cork size), and (49 – 55) mm for long aging and expensive wines. A profile of the bottle's opening is shown in the figure:

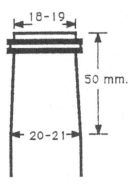

Wine bottle opening (shape and dimensions)

The dimensions of sparkling wine corks are 30 mm diameter, and about 50 mm length. Sparkling wine corks are generally made of granulated cork particles glued together (called the "body"), and one or two cork discs (about 5 mm width) at each side of the "body." With this structure the cork is capable of securing and holding high pressure of about (5 – 6) atmospheres in the bottle. On corking, the cork is compressed by the machine's jaws from 30 mm to about (15 – 16) mm (half of its diameter) and then pushed into the bottleneck, where it immediately tries to expand back to its original size. Because the inside neck diameter is between (18 – 21) mm, it remains compressed to about (60 – 70) % of its original diameter, along the tapered bottleneck. In such a compressed state, the cork adheres tightly to the glass, preventing the sparkling wine from leaking.

Back to still wines, after corking, theoretically the cork may immediately re-gain about 95% of its original diameter, and in a few hours recover to its full size. In practice, the maximum recovery is restricted by the bottleneck (about 75% of the cork diameter). Thus, the cork firmly adheres to the bottleneck. The insertion of the cork into the bottle acts as a piston, and air is compressed into it. If the newly corked bottles are placed in the down position soon after bottling, the air pressure within the bottle (up to 2-3 atmospheres) and the partially recovered cork may cause the wine to leak. As a consequence, an upright position of at least several minutes, or better some hours, is recommended. Two methods can be applied in order to reduce the pressure in the bottle. The best is to cause a vacuum in the bottle headspace at corking, so that air is not compressed during corking. The second, and cheaper, method (when such a vacuum corking machine is not available) is to inject CO_2 into the headspace before the cork is pushed into the bottle. The dissolving of the compressed CO_2 in the wine will cause a considerable reduction of the pressure. This method has a disadvantage in that the dissolved CO_2 may be slightly felt in the wine.

One should also remember that if the wine is bottled cold (white wine), pressure will build up in the bottle when the wine reaches room temperature.

From basic gas laws the new volume of a gas after its temperature was changed is: $V_t = V_o (1 + \gamma \Delta t)$

Where: V_o is the original volume; V_t is the new volume, and γ is the expansion constant, namely, the partial expansion per each extra 1°C change, and Δt is the temperature difference. Hence, the change in wine volume caused by a temperature change of Δt is: $\Delta V = V_t - V_o = V_o \gamma \Delta t$

The expansion constant for water is $\gamma = 2 \times 10^{-4}$ cc/°C.

For 10% alcohol solution it is close to 3.1×10^{-4} cc/°C.

(γ of the glass is negligible in comparison with wine; $\gamma_{glass}= 0.3 \times 10^{-4}$ cc/^{o}C). Based on these data it is easy to calculate the expansion volume ΔV at each temperature change. For example the wine in a 750 ml bottle will expand about 2.3 cc for an increase of 10 ^{o}C.

But because the volume of a closed bottle is fixed, the result will be a pressure change within the bottle according to the gas law $P_t V_t = P_o V_o$; where P_t and V_t are the new pressure and volume at the new temperature t; and P_o and V_o are the original pressure and volume, before the temperature change. The following figure demonstrates the pressure that will be built up in various headspaces (in mm) between the cork's bottom and the wine:

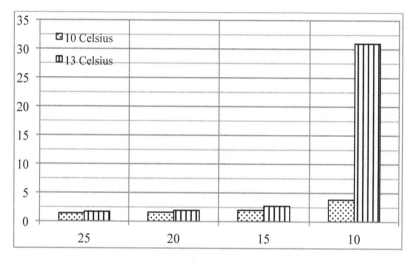

The pressure (Y- axis in atm.) that will built up upon various headspaces (X- axis in mm) at 10°C and 13°C of wine temperature increase.

If for example, only 10 mm of headspace is left at bottling (a volume of about 3.2 cc) and the storage temperature increases by 10°C, a pressure of 3.8 atm. will built up. This pressure might push (or not) the cork out, but if the temperature is a little bit higher, it will certainly do (as can be seen in the figure). Therefore it should be remembered to leave at least (2.0 – 2.5) cm length of headspace, a volume of about (6.5 – 8.0) cc in the bottleneck in order to be on the safe side.

Corks always contain dust, which may enter into the wine on bottling and can be seen mainly in white wines. Also, dried corks are difficult to slide through the corking jaws, and they tend to release more dust or small particles when bottled. It is therefore recommended to wash the corks for a very short time (a few minutes) with a solution of about 1000 ppm sulfur-dioxide, just before bottling. After rinsing, it is best to let them to dry for an hour. By doing so, most of the dust will be removed, and the slightly wet corks will more easily slide into the bottle. If the washing is done for a longer time, and the corks are more saturated, when they are squeezed by the corking machine the corks release a brown extract containing cork-tannin into the bottle, which may affect the taste of white wine.

Substitutes for Natural Cork

Good wine is always associated with a good cork stopper. But since they are a natural product, the variability of the qualities and performance of corks has brought the need to look for other alternatives that may be more uniform and predictable. A few such options are now on the market, namely, the *composite* and *synthetic* corks, *crown cap*, and *glass* corks. The composite cork is made of agglomerated cork particles bound together with food-grade glue to form a uniform cork. In general, there are fewer leakage problems compared with natural corks, but the elasticity of composite corks is lower, which may create some difficulties in pulling them out. Such corks are very much lower in price.

The synthetic corks are made of plastic materials (for example, ethylene vinylacetate), and their surface is coated with silicone oil. The materials used comply with the FDA food regulations. The plastic corks are very much denser than natural corks. A typical such cork weighs about (7 – 8) grams, compared to (3 – 4) grams for natural ones. Their permeability for liquids and gases is practically zero. The corks do not contain as many volatile compounds as natural corks, and they are not susceptible to the growth of microorganisms. Hence, no "cork taint" is expected to occur with synthetic corks. They are very uniform in structure and size; they do not break, dry out or leak, unlike natural corks. On corking, there is no dust problem with synthetic corks, and no washing is needed before use. Corking and uncorking are quite smooth, with no breakage or fragmentation. The price is quite reasonable compared to natural corks. However, the use of synthetic corks is still limited, mainly because of an image perception.

Glass stoppers have recently penetrated the market, mainly for white wines in Germany and Austria. The stopper is attached to the bottle opening by a round piece of hard plastic and the closure is verified with a "click." The capsule holds it in place. The cork is completely inert to any reaction, and so far it looks very aesthetic, with no leakage problems.

3. Capsules

Capsules are used to cover the cork surface, and serve three functions: protecting the cork from cork-borers; improving the bottle's appearance; and providing brand identity or authenticity. Traditionally, they have been made either of lead or wax. Nowadays they are made of tin foil, aluminum, or plastic. The plastic capsules are made either with thick walls and are fitted to the bottle neck by machine, or with thin walls and are shrunk under heat to fit tightly around the neck. The lead capsules were legally banned and were replaced by tin capsules which have almost the same appearance, without any health risk. The capsules are fitted onto the bottle neck by mechanical shrinking. The winery's logo is usually printed on the capsule.

4. Label

The label on the bottle is the front window of the wine in the bottle. It must represent the winery and the wine in such a way that the prospective customer will choose this bottle of wine from the full selection he/she is offered in the shop. The label may show the following important information about the wine in the bottle:

(1) The kind of wine, namely, grape variety or its name in a generic wine.

(2) The name of the winery where the wine was produced.

(3) The geographical location of the vineyard, namely, its appellation (including the state, the county, the specific local village or specific vineyard). In each state there are specific rules concerning these issues.

(4) The vintage.

These four topics are the very basic details on the wine identity. In most cases they are presented on the front label. There is additional information that is found on the wine label, partly *informal*, such as:

(a) The classified name of the wine, which the winery selects to indicate the quality of its production (e.g., basic quality, higher quality, and super quality).

(b) Regional quality status (e.g., Grand cru, Chianti Classico, etc.), which is a general classification of the winery and not a specific quality description of the wine in that bottle.

(c) Other information which the winery selects to have on the label such as winery logo, *mis en bouteille au Chateau*, unfiltered, winemaker signature, and any other detail.

And some additional information which is *enforced* by state laws, such as:

(d) The alcohol content in (v/v) as well as the wine volume.

(e) Legal warnings. In the United States, as of January 1987, the label must note that the wine *contains sulfur dioxide* if there is more than 10 ppm of total SO_2. Also other warnings in regard to health hazards have to be stated. Generally **d** and **e** are placed on the back label.

Some wineries also put additional information on the back label about the wine. This information includes some "creative" information such as barrel aging, some history, culinary/wine recommendations, and more. The front and back labels have to be approved by the authorities in most countries before the wine carrying that label can be released.

B. Bottling Line

This is final operation of wine production. It is important to remember that this last operation is irreversible—if some detail in the wine production was not done correctly, it will be impossible to correct it after bottling. The wine should be finished, stabilized in all aspects, filtered and sulfited, without any off-smell, off-taste or any other problem. All bottling materials must be ready (the right bottles, corks, capsules, labels, cardboard cases).

A flow chart of the bottling line is shown at the top of the next page:

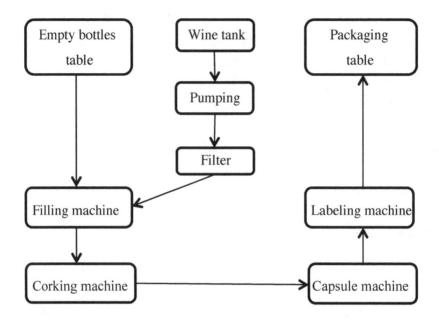

The bottling line assembly contains: an empty bottles table, a bottle washing and sterilizing device (optional), a pump and filter to transfer the wine from the tank into the filling machine, a filling machine, a corking machine, a capsule machine, a labeling machine, a packaging table to pack the bottles in cases, and a loading place for the wine cases. All the bottling elements must be connected with conveyors, so that the bottles can move easily on the line. In big wineries, the set is operated automatically, and only general supervision is needed to overcome occasional disorders in the bottling line. In smaller wineries, the system is generally not fully automatic, and workers must be present to feed the various elements entering the line, such as bottles, corks, capsules, labels, and packing cardboard. In this case a minimum of four people is needed to run the line properly. The bottling machines that come in contact with the wine (filter, filling machine, corking machine, pumps, and hoses) must be clean and sterile. Some useful points regarding bottling are mentioned here:

(1) Using the washing machine is optional. It is not necessary if the bottles were produced shortly before use (on coming from their production line they are clean and sterile), but if some months have passed, they may contain dust, insects and other materials, so it may be necessary to wash them before use. To avoid this, one should get the bottles a short time before bottling, and one should insist on buying only freshly produced bottles. If you do have to wash the bottles, the washing device contains water jets followed by air jets to dry the bottles quickly after washing.

(2) From the empty bottles table, they are fed (manually or automatically on a conveyor) into the filling machine, which usually operates (6 – 12) filling stations in small machines, and up to (36 – 72) in a large one. When a bottle in the filling machine is full, various devices controlling the liquid level in the bottle to halt the filling. The filling machine is fed with wine which flows from the tank through membrane filter of 0.45μ for white wine fill, or a coarser filter for red wine. In white wine bottling, it is advisable to place two membrane filters in parallel, where only one is in operation and the other one is used if and when the pressure on the first filter becomes higher than 40 psi (above 2 – 3 atmospheres). When wine is flowing through the filter, the wines' particles are trapped on it, and the pressure may rise. The normal pressure difference across the membrane filters when clean water is flowing is about (0.5 – 1.0) atmospheres. When pressure rises close to 3 atmospheres, the second filter should replace the first one without interrupting the bottling process. Then the first filter must be cleaned by flushing with water in the reverse direction. Be prepared to replace the other one when needed.

(3) It is not advisable to use a pulse pump to filter the wine, because of the pulsed pressure being exerted on the filter. Good filtration should be done by a smooth and steady flow of the liquid.

(4) The next stage after filling is corking. This is done by a machine which has "jaws" that contract the cork diameter before it is pushed into the bottle. The contraction should be done on the whole round surface side of the cork with even pressure; otherwise the cork may be damaged and tiny chips of cork will be released into the bottle. The cork should be pushed down into the bottle exactly to the external level of the bottle opening. After the cork is pushed into the neck of the bottle, it tries to expand back to its original size. But the neck walls block the cork, and it is held firmly by the high friction constant of the cork surface and by the conical shape of the bottleneck.

(5) The headspace between the wine level and the cork's bottom should be about (20 – 25) mm, not less, because of potential temperature expansion of the wine, and preferably not more, because of excess oxygen in the headspace. Too large space between the cork and the wine should be avoided so that not much air could be trapped in the bottle. Thus, (2.0 – 2.5) cm of headspace is recommended.

(6) During corking, pressure may develop in the bottle (because the cork is pushed in like a piston), which can rise up to a few atmospheres. It is preferable to have a corking machine that creates a vacuum in the bottle before pushing the cork

in. In the case where the corking is not operated under a vacuum, and if the bottles are placed head down (as is the usual way to pack and store wine cases), there is a good chance that some of the bottles will leak unless the pressure in the bottle is released. In such a case, leaving the bottles in the *up* position for few hours or a day is recommended, so that the pressure can decrease before the bottles are placed in the cases in the regular *down* position.

(**7**) The labeling machine is usually the "bottleneck" of the line. This machine causes most of the problems in the bottling line. Buying a good machine with reliable service is recommended. It can apply three kinds of labels: the front label, the back label and the neck label. If one decides to use a neck label on bottles, one should know that this label causes many difficulties in labeling, because of the tapered shape and small diameter of the neck. The shoulder label is the worst in this connection.

(**8**) If there is any problem in the bottling line, and the bottling has to be stopped, the first machine in the line, namely, the filling machine should be stopped, regardless of where the problem is. Also, after each station in the bottling line, it is recommended to leave some empty space for bottles to accumulate in case of some problem that may happen at the next station.

(**9**) The bottling machine and the membrane filter must be sterilized every working day before and after the filling. Passing hot water at 80°C through the filter and the filling machine for ten to fifteen minutes will do it. At the start of bottling, the first few bottles may contain some water and should be rejected from the line.

(**10**) If there is a need to touch the filling machine or the corking machine in order to fix some mechanical problem during bottling, the area should be sprayed with a 70% alcohol solution from a spray bottle after the problem has been fixed. The spraying can also be done from time to time without any specific reason, just to ensure that those parts of the machines that come in contact with the wine are sterile.

CHAPTER VII

SPECIFIC SUBJECTS

Le laboratoire de distillateur liquoriste, d'après Demachy (1775).

CHAPTER VII: SPECIAL SUBJECTS

A. Phenolic Compounds

1. Wine Phenolic

Phenolic compounds are one of the most important attributes of wine. They are responsible for wine's color, astringency, bitterness, and partially for its taste. They also play an important part in wine aging. Their chemical characteristics are very sensitive to many factors in the wine medium, and consequently they are in a state of continuous chemical changes. The basic knowledge about the phenolic compounds will be presented in this chapter. This group of compounds is called, in short, *phenolic*.

In order to be familiar with the phenolic structure and the nomenclature used in the literature of winemaking, we bring here some basic information and concepts from the general area of plant phenolic, while our attention is mostly focused on the winemaking domain. The group got its name from the phenol compound, which has no connection with this group other than its basic ring frame.

The phenol ring may be substituted by different functional groups at different carbon positions (2 to 6) to form many compounds, which are described in this chapter.

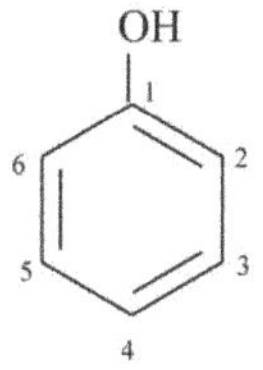

phenol

All phenolic compounds contain at least one phenol section in their structure, namely, an aromatic ring and -OH attached to it. The many phenolic compounds can be divided into three sub-groups based on their structural similarity. They are presented briefly as follow:

	Symbol	Group name
group (a)	C_6 - C_1	*p-hydroxy-benzoic*-acid
group (b)	C_6 - C_3	*cinnamic*-acid
group (c)	C_6 - C_3 - C_6	*flavonoid*

The symbol $C_6 - C_1$ represents a chemical structure of a phenolic group containing an aromatic ring (6- carbons) plus 1-carbon derivative.

The same meaning and notation relates to the other two groups, namely;

$C_6 - C_3$ represents an aromatic ring plus 3-carbon derivative;

and $C_6 - C_3 - C_6$ stands for two aromatic rings connected by 3 carbon chain. Groups $C_6 - C_1$ and $C_6 - C_3$ together are called *non-flavonoid* phenolic in order to distinguish them from the $C_6 - C_3 - C_6$ group, which is called the *flavonoid* phenolic:

Non-flavonoid	flavonoid
$C_6 - C_1$	$C_6 - C_3 - C_6$
$C_6 - C_3$	

These three phenolic groups, two *non-flavonoid* and one *flavonoid*, are the major phenolic compounds in wine and in fruits of other plants as well.

Non-flavonoid: $C_6 - C_1$

This group has the *p-hydroxybenzoic-acid* frame, where R_1 and R_2 denote various derivatives. Some of them are shown in the table:

Compound	R_1	R_2
p-hydroxybenzoic acid	H	H
p- pyrocatechuic acid	H	OH
gallic acid	OH	OH
vanillic acid	H	$O-CH_3$
syringic acid	$O-CH_3$	$O-CH_3$

Note the first compound in the table, p-hydroxybenzoic acid, upon which the whole group is named.

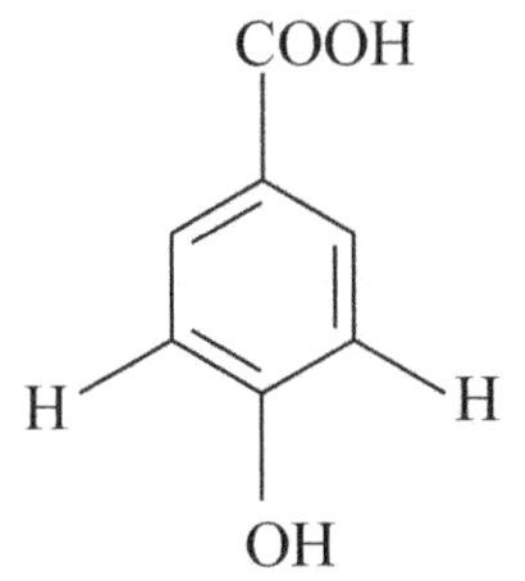

$$COOH$$

p-hydroxybenzoic acid

Non-flavonoid: $C_6 - C_3$

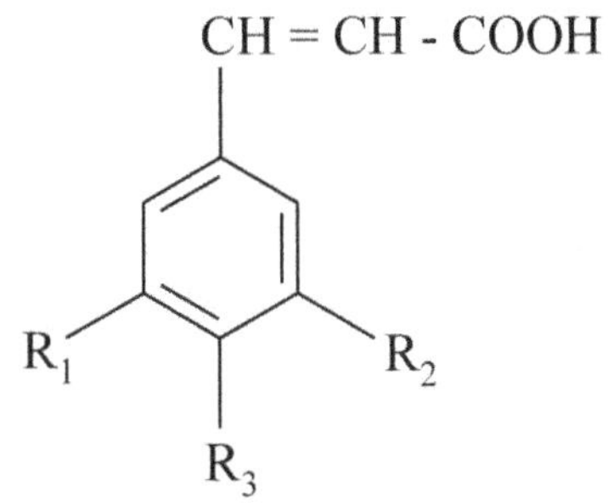

$$CH = CH - COOH$$

This group has the *cinnamic-acid* frame with three derivatives R_1 ; R_2 ; R_3 which represent the following compounds:

Compound	R_1	R_2	R_3
cinnamic acid	H ·	H	H
p-coumaric acid	H	H	OH
caffeic acid	H	OH	OH
ferulic acid	H	O-CH$_3$	OH
synaptic acid	O-CH$_3$	O-CH$_3$	OH

Note the first compound in the table, cinnamic-acid, upon which the whole group is named. This acid is formally not a phenol derivative (it has no OH connected to the ring.

181

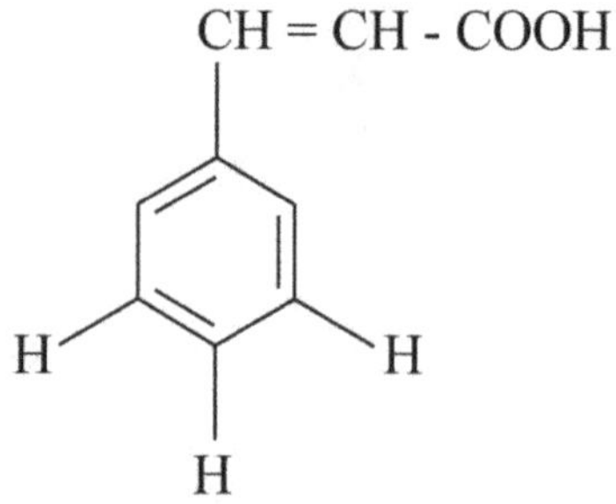

$$CH = CH - COOH$$

Cinnamic acid

Both groups of *hydroxy-benzoic* ($C_6 - C_1$) and *cinnamic-acids* ($C_6 - C_3$) are very rarely found in their acidic form (as they are presented here). Instead, they are usually connected through an esteric bond to alcohols or sugar molecules. For example, the three major compounds of this kind that are the most abundant in white wine are given here:

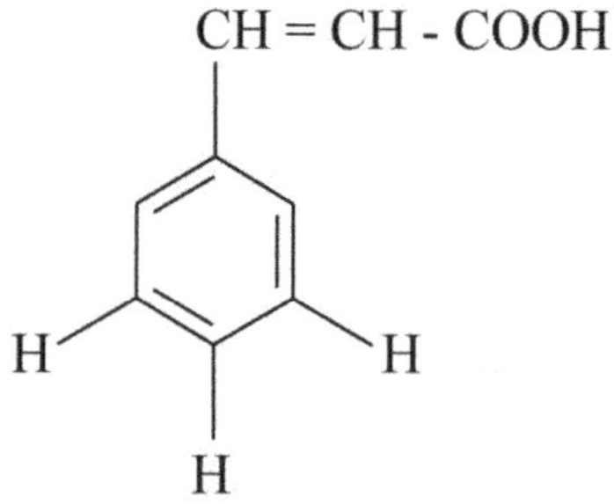

$$CH = CH - COOH$$

Cinnamic-acid derivative / tartaric acid ester

Cinnamic derivate	R_1	R_2	Compound
p-coumaric acid	H	H	coutaric acid
caffeic acid	OH	H	caftaric acid
ferulic acid	H	$O-CH_3$	fertaric acid

Flavonoid: $C_6 - C_3 - C_6$

This is the third phenolic group, the *flavonoid*, which includes the whole class of plant phenolic which has the following frame:

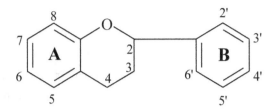

The flavonoids are structured from two phenolic rings (**A** and **B**) connected by three-carbon chain, which in most cases is closed by oxygen, constructing a heterocyclic ring (ring which includes atoms other than carbon). In some cases the three-carbon chain is open (a subgroup named *chalcones*). The numbered positions on rings A and B are the assigned sites where different derivatives are attached to the frame. The following sub-groups are listed below:

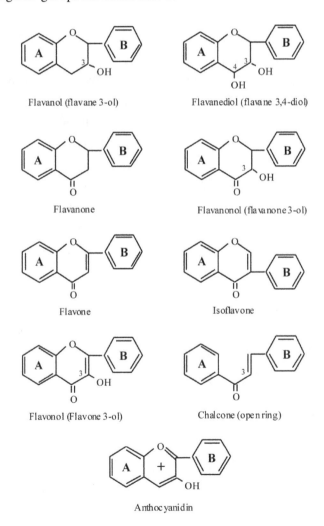

Flavanol (flavane 3-ol)

Flavanediol (flavane 3,4-diol)

Flavanone

Flavanonol (flavanone 3-ol)

Flavone

Isoflavone

Flavonol (Flavone 3-ol)

Chalcone (open ring)

Anthocyanidin

The characteristics of the different flavonoid sub-groups are derived mainly by the center heterocyclic ring, which classifies the nine sub-groups. (Note the name changing from f*lavane* --> *flavanone* when carbon-4 becomes a carbonyl, and also *flavanone* --> *flavone* when there is a double-bond between carbons 2-3.)

The differences between the many flavonoid compounds that were found in plants and in wines are based on the center ring structure and on the type of derivatives on the **B** ring. Just to mention, two of the most important are the *catechin* and *gallocatechin*, which relate to the first member on top left in the above list, the *flavanol-3*:

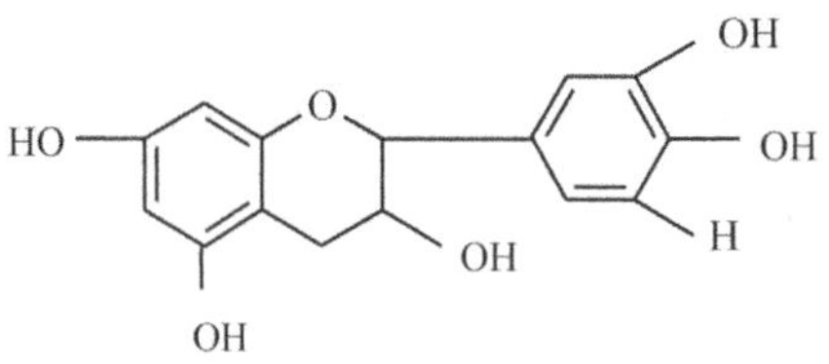

catechin 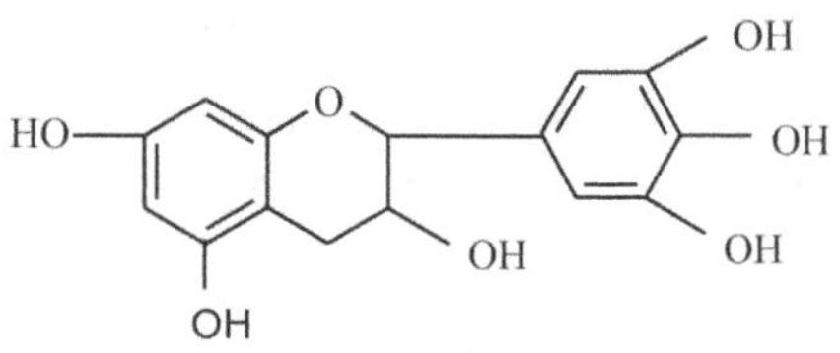   gallocatechin

The last sub-group (anthocyanidin) of the 9 flavonoids shown in the table above is a *flavanol-3* with a double bond between carbons 3 – 4 on the center ring, and which has lost one electron on this ring. As a result, this sub-group is unique among the others by being *positively charged*, and it is the basis for a large number of colored phenolic compounds called *anthocyanidin*. Among these compounds are the colored agents in wine and in other fruits as well:

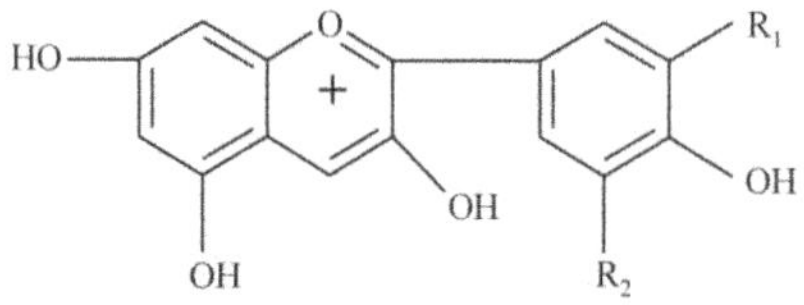

anthocyanidin

Compound	R_1	R_2
Pelargonidin	H	H
cyanidin	OH	H
delphinidin	OH	OH
peonidin	O-CH$_3$	H
petunidin	O-CH$_3$	OH
malvidin	O-CH$_3$	O-CH$_3$

The compounds listed above are the main flavonoids in grapes in various concentration ratios, and are responsible for the red color in red wines.

The phenolic compounds (*non-flavonoids* and *flavonoids*) in grapes are almost entirely located in the skins and the seeds (although the juice contains some small quantity of non-flavonoids). In white wine production some of them are extracted from the skin during crushing and pressing, or by some skin contact. In red wines, phenolic are extracted from the skin by maceration during must fermentation. The total quantity of phenolic in finished red wine is in the range of (1000 – 2000) mg/L as GAE (Gallic Acid Equivalent. See gallic-acid above among the C_6 - C_1 group). Most of them are flavonoids. Rosé wines may contain total phenolic of about (400 – 800) mg/L, half of which are flavonoids. White wine contains about (100 – 300) mg/L of phenolic, mostly non-flavonoids, but also some small amount of flavonoids, especially catechin. Oak phenolic, which are non-flavonoids (mostly gallic-acid and ellagic-acid derivatives), are contained in red wines by extraction due to barrel aging, in a range of traces up to 200 mg/L.

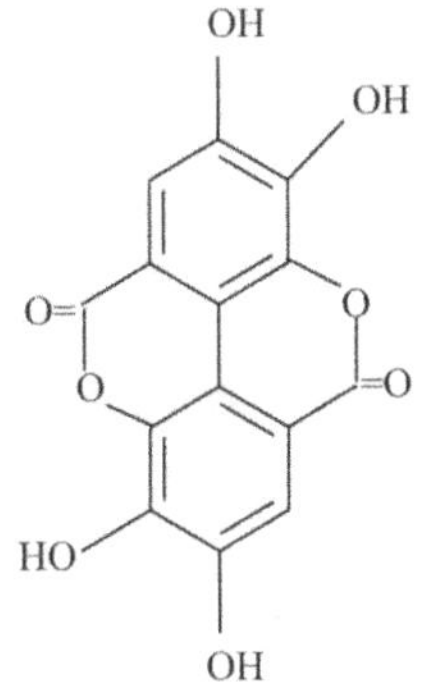

elagic-acid 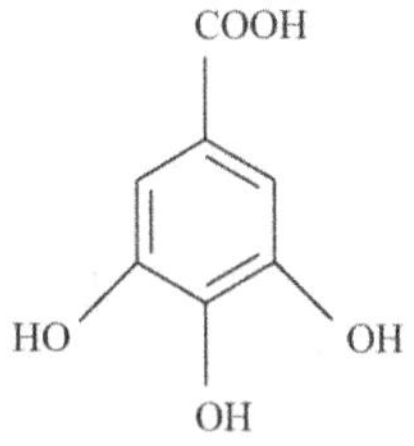   galic-acid

Glycosides

Most phenolic are linked to some other moiety and in particular to sugar molecules. The bond C - O - C is between carbons carrying hydroxyls in both the phenolic molecule and in the sugar molecules. The preferred position in the flavonoid molecule is through carbon-3 in the center ring (and also carbon-5 and carbon-7 in the A-ring). From the sugar point of view, the bond is carried on through its carbon-1 (next to the ring's oxygen), with a β-configuration (meaning the bond is above the plane of the sugar frame).

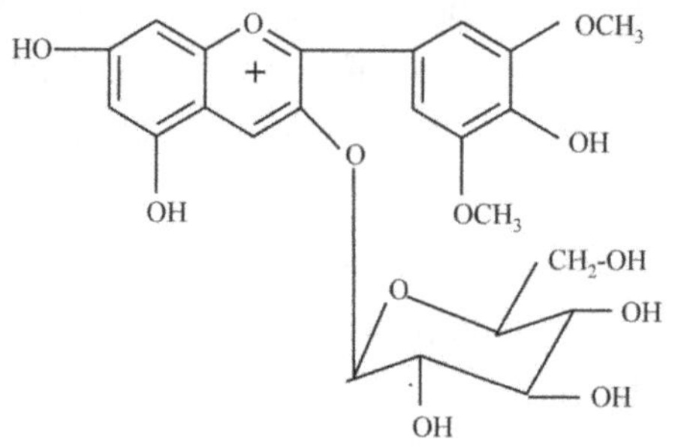

malvidin-3-glucoside

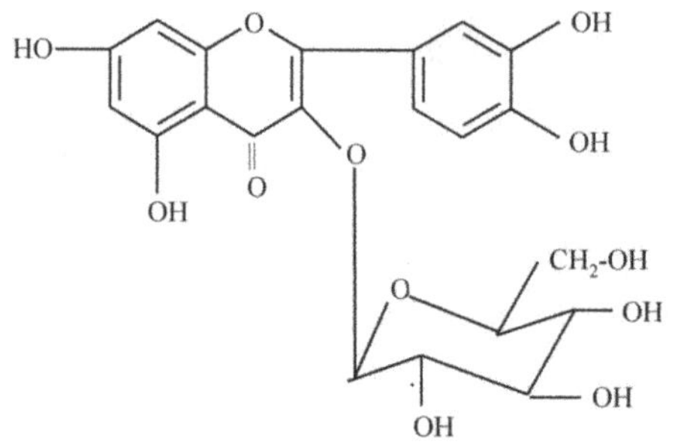

quercetin-3-glucoside

(Notice the above remarks on the bond sites between the flavonoid and the glucose.)

This bond is called *β-glycosidic bond*, and the combined molecules are called a *glycoside*. Such glycosidic bonds are very common in plants in various kinds of compounds, and in particular almost any flavor compounds in wines are glycosides. D-glucose is the most favored sugar molecule that takes part in the glycoside bond, but glycosides with L-rhamnose, L-arabinose and D-galactose are also common. Glycosidic bonding of flavonoids increases their solubility in water and enhances their stability against enzymatic oxidation.

When free *anthocyanidin* compound is glycolyzed it is called *anthocyanin*.

Also it is very interesting to note that *Vitis-Vinifera* grape varieties always contain mono-glycoside (namely, one sugar molecule per one anthocyanidin molecule), whereas native American varieties (and their hybrids too) always contain di-glycoside (two sugar molecule per one molecule of anthocyanidin). This fact can be used to distinguish between hybrids and pure Vitis-Vinifera vines in case of doubt. An example of a di-glycoside is given here:

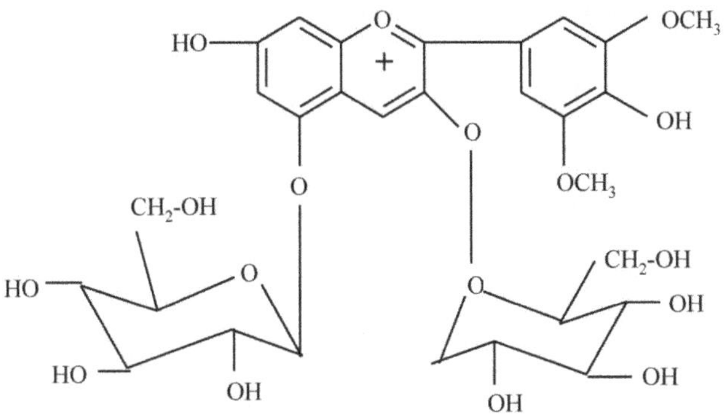

malvidin-3,5-diglicoside

The free flavonoids (non-glycoside) are also called aglycones. Acidic hydrolysis of phenolic-glycosides releases the sugar molecule and the relevant aglycone. Upon hydrolysis, the above anthocyanin (malvidin-3,5-diglucosid) will release two glucose molecules and one aglycone molecule of anthocyanidin.

2. Tannins

Historically the origin of the term *tannin* is in the leather industry, where plant extractions containing polyphenol substances were used to produce leather by reaction with hide. In solution, tannins interact with proteins, and they mutually precipitate. In leather production, the hydrogen bond cross-linkage between the polyphenol hydroxyl group and the collagen of the hide creates stability and resistance towards water, heat, and microbial spoilage. The product of such interaction is commercial leather. In order for a tannin-protein interaction to occur, it needs the suitable structure of high molecular weight tannin. The optimum size range is about (1000 – 3000) dalton, a polymer of (3 – 10) monomers. In this range of polymer size, it is big enough to create minimum cross-linkages with the collagen chain, but still small enough to penetrate into the interfibrillar region to reach the active sites.

Nowadays this term (tannin) is used as a general name for polymers of the vegetable phenolic components which have certain chemical and physical properties, such as water solubility, molecular weight in the (500 – 3000) range, and the ability to interact with proteins.

Based on their ability to be hydrolyzed, tannins were classified into two categories: *hydrolyzable* and *condensed* tannins.

Hydrolyzable tannins – these are copolymers of *gallic* and/or *ellagic* acids with sugar (in most cases glucose). The polymeric link is between the carbon-oxygen bond, and therefore it can easily be broken (hydrolyzed). These tannins are also called *gallotannins* and *ellagotannins* respectively. The exact structure of the polymers is not known but by partial hydrolysis one can find segments such as:

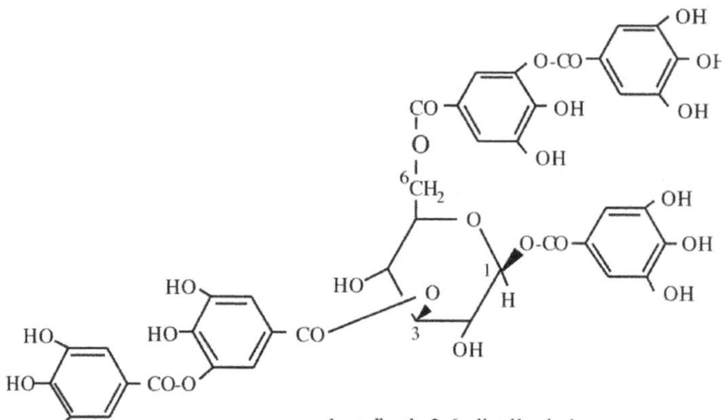

1 galloyl 3,6 digalloyl glucose

(Note the sugar molecule in the middle with connections to three different phenolic.) Such tannin is easily hydrolyzed to the sugar and phenolic components. Commercial "tannic acid" is a mixture of such glucose-gallic acid polymers.

Condensed tannins – these are polymers of flavonoids which are condensed mainly through a carbon-carbon bond. Such bonds are not easily hydrolyzed under regular conditions. The polymerization between the monomers takes place through carbons 4 ; 6 ; 8 ; in the flavonoid frame. The most common condensed tannins are polymers of flavanol-3 (catechin group and their epi-isomers) or the less common flavone-diol 3,4 (leucocyanidin group). They are called also *procyanidins* or *leucocyanidins* because under heat and acidic conditions they may dissociate.

The classification of tannins to *hydrolyzable* and *condensed* has only historical meaning. Both classes can be hydrolyzed. So the use of these terms is more to differentiate between the main monomers composing the polymers: gallic-ellagic-glucose is considered the *"hydrolyzable tannins,"* and flavonols are the *"condensed tannins."*

Some examples of the condensed tannins (two trimmers and one tetramer) are shown at the top of the next page:

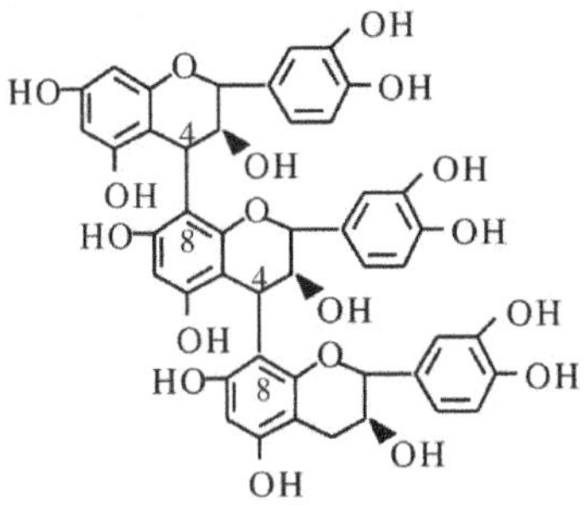

4-8, 4-8 tricatechin

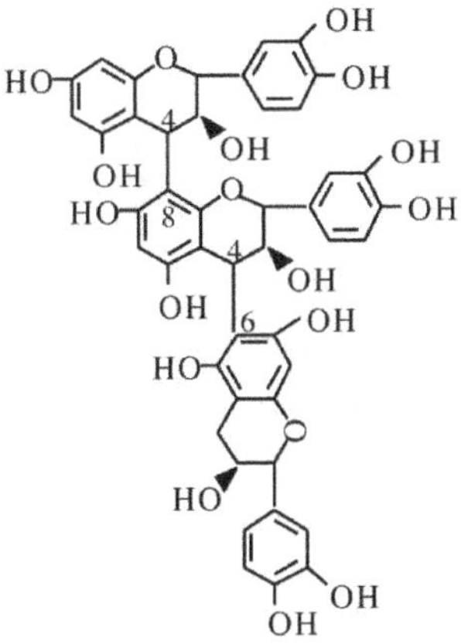

4-8, 4-6 tricatechin

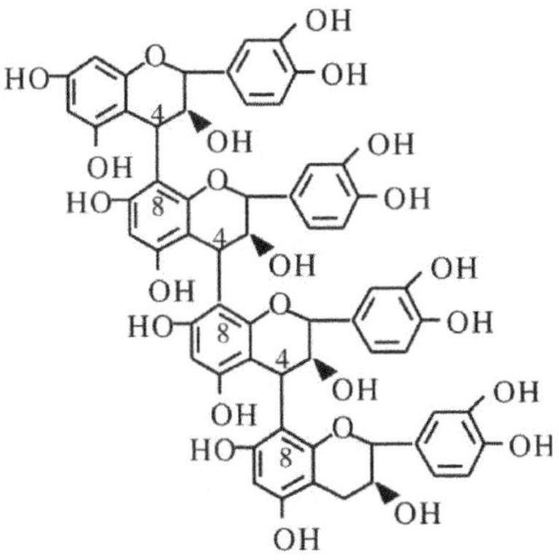

4-8 all tetracatechin

Combinations of a mixed flavonoid and non-flavonoid polymers are also known. One such copolymer between gallocatechin and gallic-acid is shown:

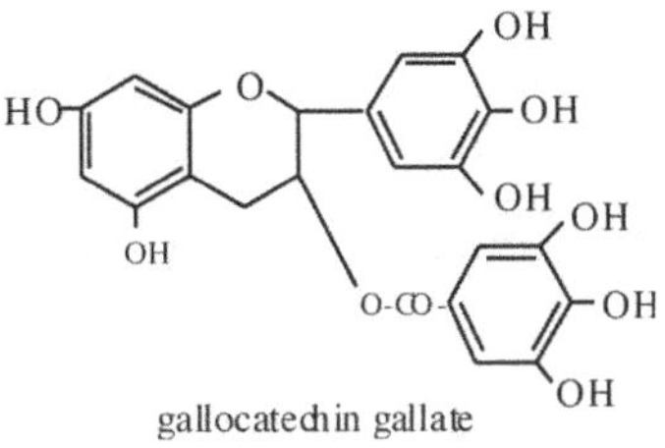

gallocatechin gallate

The molecular weight of a monomer is about 300, and the number of units found in wine's tannins is up to about 10. Hence the molecular weight of wine's tannins is between (500 – 3000). The relative distribution of tannins in wine, with respect to their polymerization size, changes during wine aging. This subject is discussed in section B of this chapter (oxidation). Also, during aging in oak barrels, hydrolyzable tannins are extracted into the wine, a process which influences the wine's taste and bouquet very much.

The content of *polymeric phenols* in free-run young white wine is less than 1% of the *total phenolic* concentration, which is about (100 – 300 mg/L). Upon skin-contact of several days it may increase to (2 – 5) % of the total in varieties such as Colombard and Sauvignon Blanc, and up to (10 – 20) % in Semillon and Chardonnay. In red free-run juice the polymeric phenols are (5 – 15) % of the total phenolic concentration and may increase to (20 – 40) % upon long skin-contact. In aged red wine the polyphenol concentration is constantly increasing. Stems and leaves contain significant concentration of polymeric phenols, up to about 10,000 mg/kg and 6,000 mg/kg respectively. However, during regular wine processing, their contribution to the phenolic concentration has no significance because they are removed from the must.

3. Red Wine Color

As was mentioned above, series of anthocyanin compounds are responsible for the red color in red wines. They are well-known pigments in plants, responsible for the wide range of colors (almost the full visible spectrum) in flowers, fruits and leaves. The representative formula of anthocyanin is:

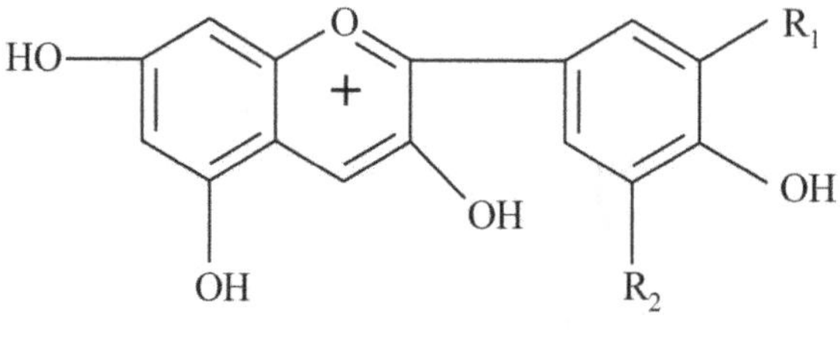

(where R_1 and R_2 are a combination of the derivatives H ; OH ; O-CH_3).

The actual color in plants depends on various parameters such as the specific anthocyanin and its concentration in the tissue cells, and on the pH in the cells. In most fruits there is one major anthocyanin which dominates its color, for example *cyanidin* (in apple, blackberry, fig, peach, red cabbage), *delphinidin* (in passion fruit, pomegranate, and eggplant), *cyanidin* and *peonidin* (in cherry, plum, and onion). In Vitis-vinifera grapes all six anthocyanidins mentioned above—*pelargonidin, cyanidin, delphinidin, peonidin, petunidin*, and *malvidin*—are represented. The last in the list, malvidin, has the highest concentration.

pH dependence of anthocyanin color

Anthocyanins in solution are involved in equilibrium with other flavonoids which may have different colors, and because protons are involved, this equilibrium is pH dependent. The following model which represents the pH dependence of red wine color on the pH was suggested. The anthocyanin red cation takes part in two sets of equilibrium:

(1) One by water molecule addition and losing its proton to form neutral *carbinol*, which is colorless and which is also in equilibrium with its yellow open ring tautomer chalcone.

(2) A second equilibrium takes place (also by losing its proton) with the anhydrobase *quinoidal* form, which is violet. The quinoidal form is also in equilibrium with its anionic tautomer (negatively charged), whose color is blue. It is clear, then, that the color and intensity in anthocyanin solution will be very much dependent on the concentration ratio of these sets of compounds whose equilibriums depend on the solution pH.

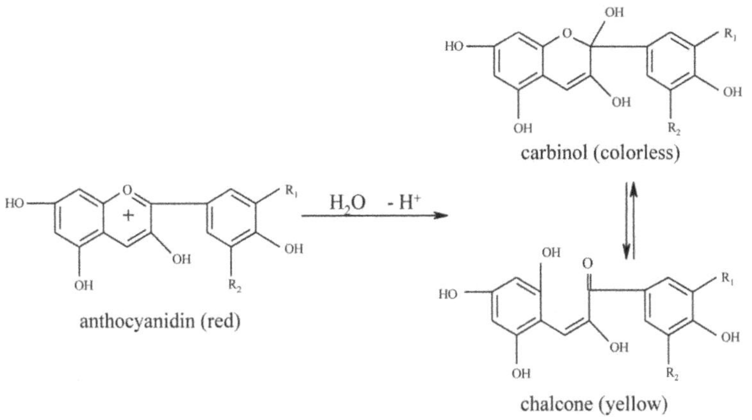

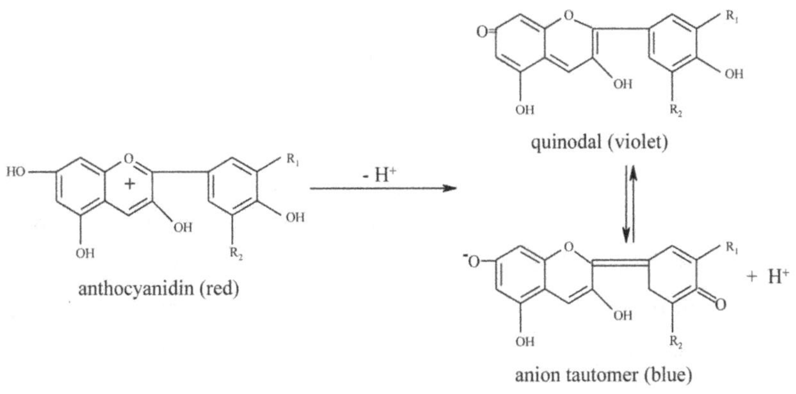

carbinol (colorless)

anthocyanidin (red)

$H_2O \quad - H^+$

chalcone (yellow)

quinodal (violet)

anthocyanidin (red)

$- H^+$

$+ H^+$

anion tautomer (blue)

To clarify, let's deal with a specific case by looking into the major component in red wine pigments *malvidin-glucoside*, whose pK_a's for the two equilibriums (at 25°C) are known to be 2.6 and 4.25 respectively. Let summarize the case of malvidin-glucoside as follows:

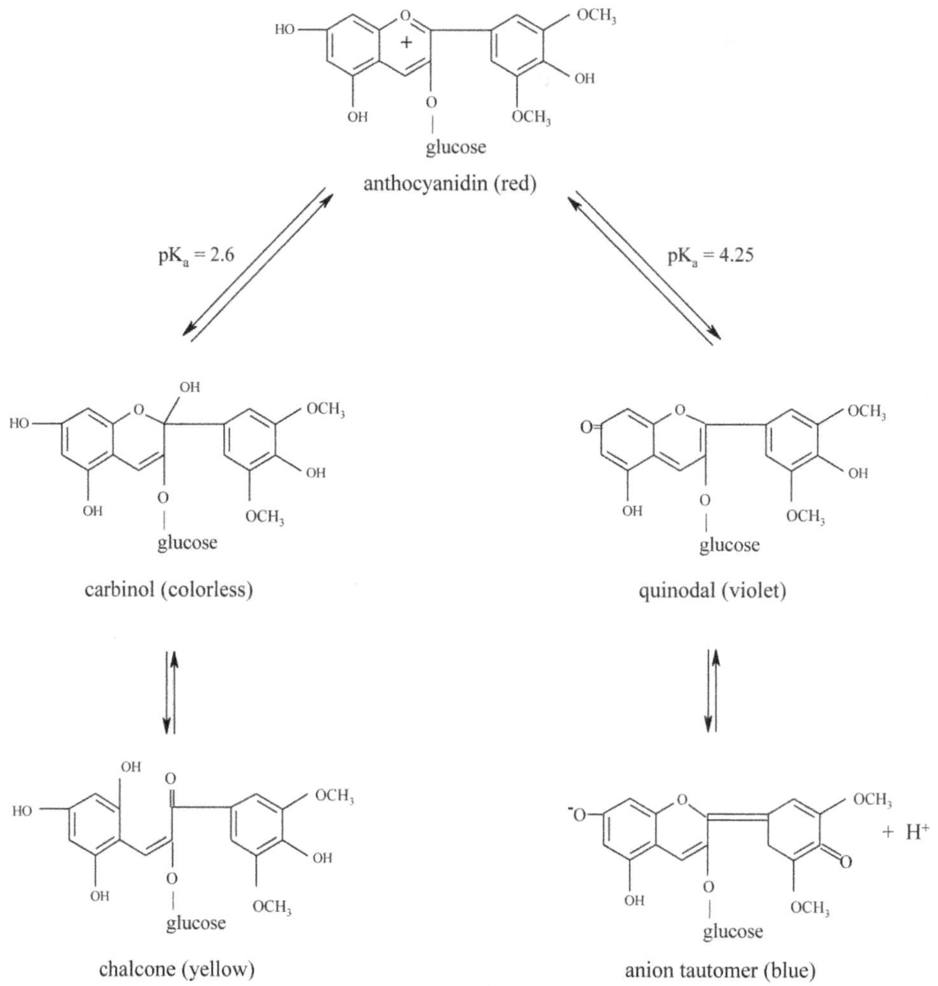
HO
OCH3
OH
O
glucose
anthocyanidin (red)
pKa = 2.6
pKa = 4.25
carbinol (colorless)
quinodal (violet)
chalcone (yellow)
anion tautomer (blue)
+ H+

Equilibrium scheme between malvidin and its other forms in solution

From these equilibrium equations, and the known relation between the various compounds in the solution, it is possible to calculate the relative concentrations of each component in solution at any pH value, and hence the color composition and intensity of the solution. (The reader who is interested in the full details is referred to the *Concepts in Wine Chemistry* book, chapter III, section C.)The relations between the components are summarized in the equation:

$$(10) \qquad \log [F^+] / [F] = pk_a - pH$$

where: $[F^+]$ - is the protonated form of malvidin in molar concentration.

$[F]$ - is any other forms of malvidin in molar concentration.

pK_a , pH - are the relevant pK_a (2.6 or 4.25) and the solution pH.

Based on the above equation, malvidin equilibrium in low pH range appears as:

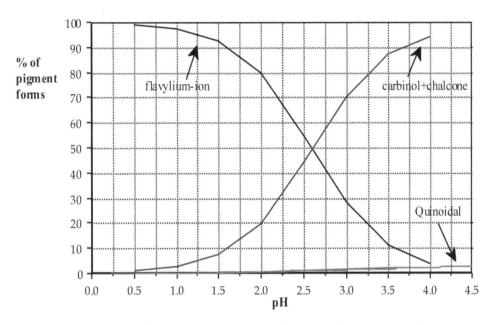

Notice how low the red colored form (malvidin-ion) is in the wine pH range of (3.0 – 4.0). For example at pH = 3.25 it is about 20%, and at pH = 3.5 only 10% of its contents is red-colored. On the other hand, it is practically 100% at pH below 0.5. This figure shows two significant facts: one, how low is the actual concentration of the red fraction of anthocyanin in red wine, and second, how sensitive is the red color intensity to the wine pH. In other words:

The lower the wine's pH, the higher the intensity of its red color.

Let us analyze some specific examples at various pH which may clear the case:

(a) Solution at pH = 1.6

This pH is far away from the right side of the above equilibrium scheme between *malvidin* and *quinoidal*, where the pK_a = 4.25 is greater by almost three orders of magnitude than the pH, and therefore this side of the equilibrium can be ignored. On the other hand, the left side of the equilibrium scheme between malvidin-carbinol can be described by using equation (10):

$$\log [F^+]/[F] = pk_a - pH = 2.6 - 1.6 = 1.0 \qquad ===> \quad [F+]/[F] = 10$$

which means that the concentration ratio malvidin to (carbinol+chalcone) = 10, and the solution color will be deeply red (ten parts of red malvidin to one part of colorless/yellow of the other two).

(b) Solution at pH = 3.6

The pH solution is between the two pK_a's, so both equilibriums must be considered.

For malvidin/carbinol: $\log [F^+]/[F] = 2.60 - 3.6 = -1.0 \ ===> [F+]/[F] = 0.1$

For malvidin/quinoidal: $\log [F^+]/[F] = 4.25 - 3.6 = 0.65 \ ===> [F+]/[F] = 4.5$

This means that the concentration ratios of the compounds at equilibrium, namely, (carbinol+chalcone) / malvidin / quinoidal are equal to **10 : 1 : 0.22**, which is about 89% (carbinol+chalcone); 9% malvidin; and 2% of quinoidal forms. Since carbinol is colorless and malvidin is red, the color at this pH is still red, although much less intense than at the above pH = 1.6. The small amount of the quinoidal form will also contribute some violet tone to the color.

(c) Solution at pH = 4.25

The (carbinol+chalcone) / malvidin / quinoidal ratios in the solution can be calculated in the same way as above (the reader is invited to do so) and found to be **95 : 2.5 : 2.5**. The solution contains only about 2.5% of red pigment and the same amount of blue. The rest is colorless; hence, the color will be a mixture of light red and blue, and tends to look red-grey.

(d) Solution at pH = 5.6

The ratio of malvidin to its components (i.e., malvidin/(carbinol+chalcone) on the left side of the equilibrium is found to be: $\log [F^+]/[F] = 2.6 - 5.6 = -3.0$ ===> $[F+]/[F] = 10^{-3}$. This means that the malvidin concentration is only 0.1% at this pH. The equilibrium is practically established between the right side of the scheme. Therefore, no red is seen, and the color is a kind of grey hue.

4. Extraction of Phenolic Compounds from Grape Skins

The various phenolic distributions in white and red wines (in mg/L as Gallic acid Equivalent) are:

Phenolic	White	Red
Non-flavonoids (mainly hydroxy-cinnamic-acid derivatives + their esters)	200	200
Anthocyanins (mainly malvidin -3 glycoside)	----	150 - 300
Non-colored flavonoids (mainly catechin and epicatechin)	25	200 - 250
Tannins (mainly polymeric catechins and others)	very low	500 - 1000
Total (mg/L)	**225**	**1050 - 1750**

The phenolic compounds in grapes are concentrated mainly in the berries' skins (about 50% of the total), in the seeds (about 45%), and the rest (5%) in the juice. The various groups are distributed as follows: Anthocyanins are located in the skins. The monomeric phenolic of the non-flavonoid groups and their polymeric tannins are found in the skins and the seeds. And the cinnamic-acid phenolic group plus its esters are mainly in the juice and in the skins. The other flavonoid compounds (colorless compounds) are also mainly in the skins and in seeds. Therefore, there is a large difference in phenolic concentration in white and red wines. White wine contains a low concentration of phenolic, because it is made from juice pressed before fermentation. So only a small fraction of the phenolic found in the berries is introduced into the must.

In red wine production, the fermenting must contains skins and seeds during and after fermentation, which are extracted into the must/wine at a certain rate, and eventually, if enough time is allowed before pressing (separating the skins and the seeds) they will reach concentration equilibrium between the skin/seed cells and the outside medium. The extraction is a diffusion process which obeys basic diffusion rules:

(1) The direction and rate of diffusion are controlled by the concentration gradient of the components under study. Simply stated, this means that the components

flow from a higher concentration to a lower one, and the rate of flow is faster when the concentration difference (the gradient) is higher.

(2) The diffusion rate is controlled by the temperature. It is higher at elevated temperatures.

(3) If enough time is allowed, it will come to a dynamic equilibrium between the two phases. The phases are the skin/seed cells and the wine medium. Equilibrium will be reached only when the multiplication of two factors, the *concentration x affinity* in both phases, becomes equal. In most cases the affinity of a compound to one phase, is not the same as to the other phase. An example of two such phases is a mineral salt in water and organic solvent, where it has a very high affinity to water and almost zero affinity to the organic solvent. Water and membrane tissue of a living cell are another example. Therefore one should keep in mind that only a certain percent of the color components and tannin contained in the grape tissue will diffuse into the processed wine, actually less than 50%. The rest of the phenolic will remain in the solid must particles (skin, seeds, pulp, tartrate crystals, and even yeast cells).

The following factors determine the final composition of phenolic in wine:

Grape variety **and** *viticulture parameters*

Different varieties typically have a different phenolic content, which finds its expression in the color intensity and tannin content of their wines. "Nature," namely, the location of the vineyard (climate, soil, and elevation) and the vintage, are also very important factors in this matter. The grape-growing practice, which includes all the growing parameters, such as irrigation policy, fertilization, canopy management, and others, also determines the phenolic concentrations in a given vine. In a study measuring the tannin concentration in a very large sample of wines (1300 samples) made of many varieties and in many regions in the world, it was found that there is a big difference in phenolic concentrations between varieties on one hand, and a huge standard deviation within each variety, up to half of the mean value, on the other hand. Cabernet Sauvignon, Merlot, and Syrah are at the upper end, while Pinot Noir is at the lower end. The average tannin concentration in Cabernet Sauvignon was found to be (672 +/- 303) mg/L, and in Pinot Noir wines (350 +/- 200) mg/L as CE (Catechin Equivalent). The variability among grapes in each variety is so large that no conclusion can be made about the variety of the wine, based on its tannin concentration.

Temperature of fermentation

Temperature profile during fermentation will affect the extraction rate. As the fermentation temperature becomes higher, the extraction rate gets higher. There are three reasons for this: first the permeability of the skin cell membrane is increased at higher temperatures, allowing faster release of anthocyanin and tannin molecules outside; second, the solubility of those phenolic compounds in the wine solution increases at higher temperatures; and third, diffusion of soluble compounds is faster at higher temperatures. This issue was investigated in many studies and was confirmed.

Maceration profile

At the beginning, when the must is kept in the tank before yeast addition, the diffusion occurs solely in a water solution, where the color components (anthocyanin) are most soluble, causing their extraction rate to be faster. The rate slows down later, when the alcohol content gets higher with fermentation. The concentration gradient also becomes smaller as the liquid must/wine becomes richer with anthocyanin. At the same time, the tannins are better extracted, because the water/alcohol solution becomes more alcoholic, and their extraction rate grows slowly with the fermentation. Separation of the skins from the wine marks the end of the extraction process. The timing is therefore very crucial in this matter. Short maceration, close to the end of regular fermentation (about six to eight days) will leave the wine with almost its maximum color components, because these components are the fastest to be extracted. But the tannins at this stage are not yet at equilibrium with the wine solution; therefore an extended maceration, practiced up to (20 – 30) days beyond finishing fermentation, will increase the tannin content.

Because the skins are pushed up during fermentation by the CO_2 evolving from the fermenting must, a cap is formed on top of the must, with two main results: there is no contact between the juice and the skins, and heat cannot get out, which elevates the cap's temperature very much. The cap has to be treated to enable skin/juice contact and to release the high temperatures. This operation is done mainly by two methods, either by punching down (manual or mechanical) or by pumping-over the liquid must/wine on top of the cap to wet it and flow down through it. Each of these methods is done few times a day until skin separation.

In order to facilitate the extraction rate, especially in large-scale wineries when time is very short at harvest, a special rotating tank was developed. In these tanks, instead of using the traditional punch-down method or the pumping-over method to wet the cap and maintain the contact of the skins with the wine, the whole tank rotates (horizontally) several times a day in both directions, with spiral blades inside the tank to break the cap, so that the skins are in good contact with the fermenting wine. The operation is planned to take a few days (3 – 4), in order to save time, and then the wine is pressed and finishes the fermentation in ordinary tanks. The idea behind these expensive tanks is that in a short time (up to 4 days) it is possible by a mechanical device to extract the phenolic at the same final content as would take (8 -12) days in a regular tank. The only problem is that the final concentration depends not only on the diffusion rate between the skin/must contact but also on the dynamic equilibrium state between the two phases. Once it is established, no additional phenolic can move out from the skins. Thus, the rate of extraction may be higher with this technique, but a limit is set up by the equilibrium state which is time-dependent. In any case, no matter what technique is used, the final concentration is defined only by reaching equilibrium. The final wine made by that technique after (3 – 4) days of maceration (with good skin/juice contact) is still lower in phenolic content than one would expect, due to the shorter maceration time. In any case, the profile of anthocyanin during extraction is that it reaches its maximum content during the second to fourth days of fermentation, and then gradually decreases at all stages of processing and in the bottle. The tannins, on the other hand, slowly increase in content until pressing, and then stay stable throughout processing until bottling.

Other attempts which are used in the industry to enhance tannin content and anthocyanin in wine are:

(1) Cold soak of the must before the fermentation for a few days.

(2) Long maceration time after fermentation.

(3) Pre-heating of the whole grape cluster before fermentation (called also thermo-vinification).

(4) Freezing of the berries before fermentation.

(5) Carbonic maceration with CO_2.

(6) Pectolytic-enzyme treatment to break the cells of the skins; and lastly

(7) Juice bleeding after crushing to increase skin/juice ratio.

Many studies have been done on these techniques, some of them with contradictory results using the same technique. But the overall conclusion can be drawn that right after fermentation, the concentrations of anthocyanin or tannins in some of these treatments were higher to some extent, but afterwards in the long run, practically no advantage was observed in the finished wine, including wine quality that was measured by sensory evaluation. **The only method which practically enhances the phenolic content in the finished wines for the long run is higher temperatures of fermentation.**

B. Oxidation and Micro-oxygenation

1. Oxidation

Wine oxidation is a term given to a set of very important reactions which occur during wine processing and aging. It involves reactions between air oxygen and various substances in wine such as phenolic compounds, aldehydes, sugars, sulfur-dioxide, and others. Any time when wine (or must) is handled (crushed, racked, pumped, filtered, topped, and bottled) aeration takes place, with some amount of oxygen dissolving in it. The dissolved saturation values of oxygen in liquid are inverse temperature-dependent, namely, there is higher dissolution at lower temperatures. The saturation values for oxygen (in mg/L and cc/L) at atmospheric pressure in water at different temperatures are given in the following figure:

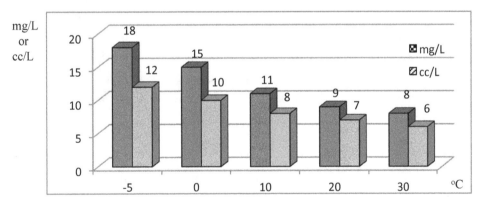

Saturation values of dissolved air-oxygen in water at various temperatures

In wine these values are about 20% lower. Dissolved oxygen and its control in wine are key factors in wine oxidation and aging processes. After the oxygen has been used up during oxidation processes, any handling of the wine which exposes it to air will be followed by a new dose of dissolved oxygen, and further oxidation. Any operation in the cellar, such as pumping wine from one tank to another, introduces about (2 – 7) mg/L of dissolved oxygen, depending on how well-protected the transfer is, and on wine temperature. The actual concentration of dissolved oxygen in wine can be easily measured by an oxygen meter probe. The fate of 5 mg/L dissolved oxygen in red wine was measured (at 20°C) and was found to steadily decrease until it stabilized after 48 hours at 0.5 mg/L. The decrease in oxygen concentration and its effect on wine is the issue in this section of the chapter.

Wine in general, and red wine in particular, is a very complicated chemical system which undergoes constant changes caused by many and various reactions. Wine aging is a term that includes all chemical changes that happen in a wine's lifetime and find their expressions in its color (from light-yellow shifting to deeper yellow in white wines, and from violet-red shifting to tawny-red in red wines), its transparency, its bouquet, and its taste. Most of these changes are the results of oxidation processes, whose products and rates depend on how much the wine was exposed to air, on its temperature history, on the wine pH, and on the measures that were taken to protect the wine during and after vinification. Experience shows that exposure of white table wine to air is in most cases detrimental to its quality (loss of fruitiness, caramelization, sherry-like odor), whereas slow exposure of red wine improves it by softening its tannins and by developing its bouquet. It seems therefore that the "buffering" potential of white wine to resist undesired oxidation is much less than that of red wine. On the other hand, intensive overexposure of red wine to air/oxygen might cause significant reduction in its quality. The major groups of components in wine that play a key role in its reaction with oxygen are its phenolic. This may help to explain why white wines are so much less tolerant to oxidation than red wines, on the basis that the average total phenolic content in red wines is about ten times higher than in white wines.

Phenolic Oxidation

There are two major mechanisms of wine oxidation, namely enzymatic and nonenzymatic. The enzymatic oxidation of phenolic compounds in plants is carried on by two types of enzyme groups:

(1) *Polyphenoloxidases* (PPO) – which have also various synonyms such as *catecholoxidase*, *tyrosinase*, *phenolase*, *cresolase*, and *o-diphenoloxidase*.

(2) *Laccase* – which has also the name *p-diphenoloxidase*.

Both enzyme groups catalyze the oxidation of plant phenolic with molecular oxygen. But the first enzymes, *polyphenoloxidases*, which are dominant in plants, are able to oxidize only part of the phenolic compounds (the *ortho-diphenols*, such as *catechol*, *caffeic-acid*, *gallic-acid*, *catechin*, *epicatechin*, and others. The *ortho* means two neighbor groups on the ring):

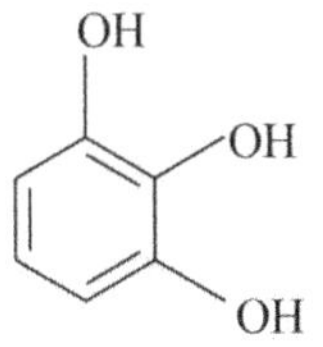

gallic acid 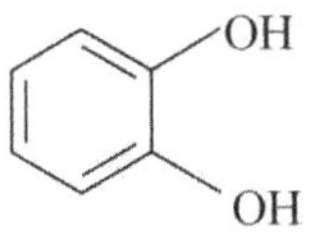   catechol

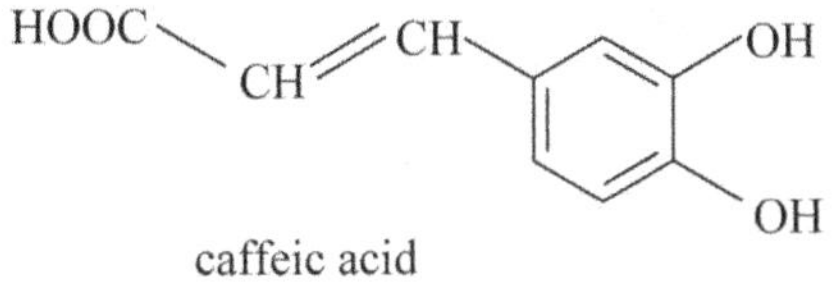

caffeic acid

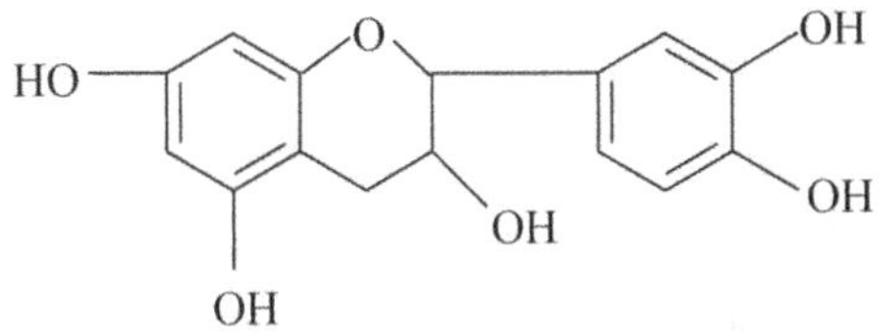

catechin

These enzymes are found in many fruits, including grapes, and are responsible for the typical browning of cut fruits (e.g., potatoes, apples, and grape must). They are very sensitive to sulfur-dioxide inhibition, which is the main reason for its use right at crushing.

The second enzyme, *laccase*, has a higher capability to oxidize a wider range of phenolic, in addition to the above ones, such as monophenols (one OH on the ring) such as:

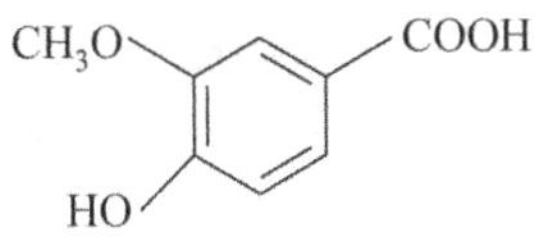

vanillic acid 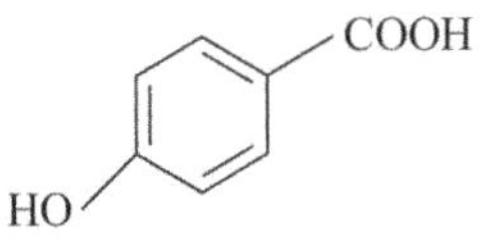   p-hydroxycinnamic acid

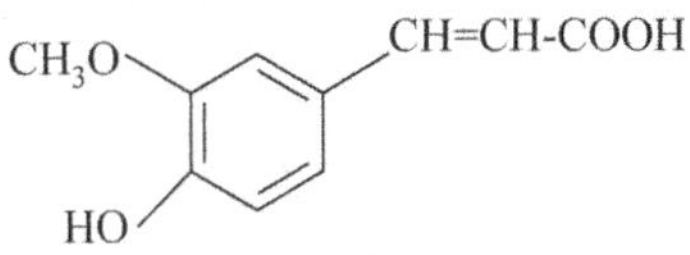

p-coumaric acid 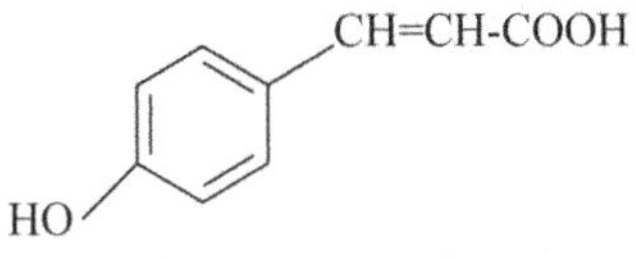   ferulic acid

Laccase is not found in healthy grapes, but it is carried into the must by the mold of infected grapes. It is quite resistant to sulfur-dioxide, stable and active in the wine medium for a long time, and may cause a lot of oxidation problems. Therefore, if it is possible, minimizing moldy grapes is highly recommended (by sorting them out before crushing). But if the crop is heavily molded (specifically by *Botrytis-Cinerea*), higher concentrations of sulfur-dioxide will be needed, as well as lower storage temperatures to slow down the enzyme activity. If possible, lowering the pH will also help to reduce the oxidation rate, an effect which will be discussed later. Another option in a severe case of moldy grapes is to flash-pasteurize the must (technically possible only on white clear must) at a temperature above 80°C in order to deactivate the enzyme. (Flash-pasteurization is a very fast heating of the flowing must or wine in a heat exchanger for very few seconds, and fast cooling back with minimal damage to the potential wine flavor.)

Back to *polyphenoloxidase* (PPO), which is the major enzyme of interest in wine production, we will focus the discussion on the oxidation caused by this enzyme. PPO, which contains copper atoms in its active site, is able to perform the following oxidation reaction on phenolates:

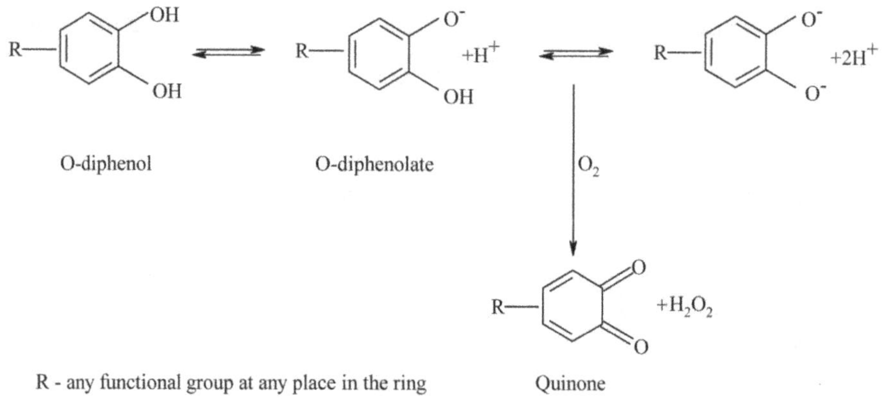

| O-diphenol | O-diphenolate | | Quinone |

R - any functional group at any place in the ring

The results of the oxidation are quinone and hydrogen-peroxide, two molecules that continue the process in different directions. The quinone is quite active and able to undergo free-radical reactions to form carbon-carbon bonds with other phenolic molecules to create a dimer, trimer, and so on. This is the phenolic polymerization pathway to produce polymeric phenols which are called *tannins*. As for the hydrogen-peroxide, it is a strong oxidizing agent which can continue the oxidation reactions, mainly with ethanol, to form acetaldehyde (see later in this section).

Red Wine Oxidation

In red wines, the anthocyanin red pigments are polymerized with other phenolic (such as catechin and various tannins) to form colored pigments which in general have more intense red color in the wine pH range. This co-pigmentation, which is a result of phenolic oxidation, is very fast during the first year in the barrel (up to 50% polymers out of the total pigment content), and may reach (80 – 90) % after ten years of aging in the bottle.

In aged wines, then, the polymeric forms of anthocyanins are the major contributors to the red wine color, rather than the monomeric anthocyanins. The polymers are more stable towards further oxidation, so the rate of aging process and the color change to brown-red are slowed down after a few years of aging. The typical pattern of color changes in red wine during its lifetime are from violet-red in very young wine, to deep red after few years of aging, followed gradually by brown-red, and eventually to brown in an over-aged wine. The violet-red color of very young wine is caused by the almost pure content of red monomeric anthocyanin, which is in equilibrium with one of its violet color components (anhydro-base quinoidal) in a small percentage (see previous section.) On oxidation this component loses its violet color, and the deeper red color of the copigments takes over. All this means that as the wine is oxidized, and consequently its anthocyanins become partially converted to polymers of anthocyanins and other phenolic, the colored fractions in the wine solution (which is pH-dependent) become higher in accordance with the oxidation process. And as a consequence, the color intensity is enhanced. From a practical point of view, aeration of a red wine tank (if barrel aging is not planned) will contribute some additional color intensity and some higher quality, as compared to an un-aerated tank. The importance of anthocyanin-tannin copigmentation in preserving light-colored red wines (e.g., Pinot Noir) is well known in Burgundy. To enhance the tannin content, many wineries add some portion of stems back into the fermenting tanks.

White Wine Oxidation

The phenolic compounds in white wines are mainly cinnamic-acid derivatives and their tartrate esters, such as caftaric-acid, coutaric-acid, and fertaric-acid

Their total content in white wines is in the range of (20 – 200) mg/L. Another group of phenolic compounds are the flavonoid (catechin, epicatechin, gallocatechin), with a total content of about (20 – 100) mg/L.

As a result of the oxidation process, polymerization takes place, and the polymers are responsible for the yellow-brown color of white wines.

The o-quinone formed by the phenolic oxidation (mentioned above), are among the main precursors of browning reactions in other fruits as well. For example, the main substrate for oxidation to o-quinone in apples and potatoes is non-flavonoid phenolic such as chlorogenic-acid, and in onions the *protocatechuic-acid*:

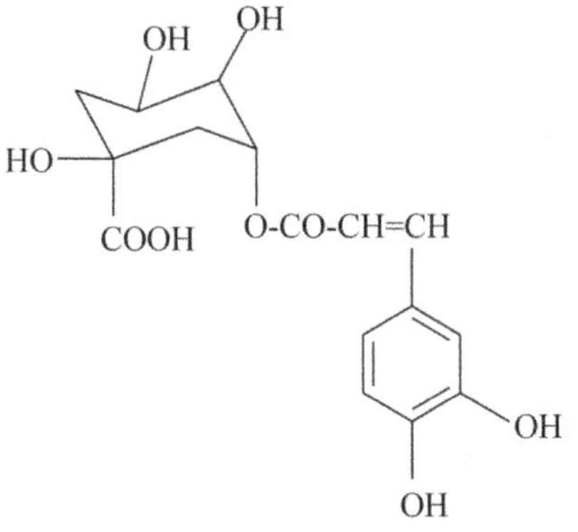

Chlorogenic acid 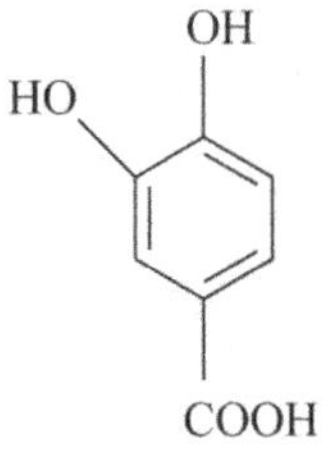   protocatechuic acid

Facing these problems of white wine oxidation, a new approach to the issue was initiated during the last two decades of the twentieth century, to oxidize the must on purpose *before fermentation*, in order to cause and enhance phenolic compounds to polymerize and to precipitate, thereby eliminating the problematic phenolic in white wine. This idea is completely in contrast to the traditional methods for protecting wine against oxidation. Three questions have to be answered in order to consider this idea: (1) will it work, namely, is it possible to eliminate the relevant phenolic from the must? Then, (2) after doing it, will the wine be protected? And last, (3) what would be the effect on wine aroma and its quality? At the right conditions, fast enzymatic oxidation will take place, and the must will turn dark-brown, and most of the phenolic that are not soluble in must, will precipitate. It is hard to believe, but the must that is almost black, after finishing fermentation, becomes very light-colored wine (the dark color particles are absorbed by the yeast or precipitate). What are the technical details of the procedure? First, not to add SO_2 to the must after crushing, and second, to supply and maintain oxygen uptake by the must in one of two alternative ways:

(1) By pumping the must from one tank to another one, or back to the same

tank, and adding the oxygen into the line by a diffuser from an oxygen cylinder through a flow rate device. The volume of must and the desired oxygen concentration can guide the oxygen flow rate; for example if the must volume is 10,000L and the desired oxygen uptake is 12mg/L, then by setting the oxygen rate to 4 gram/min for 30 minute, 4x30x1000/10000 = 12mg/L will be the oxygen uptake by the must. This procedure can be applied more than once (twice or three times), in order to complete the assumed full oxidation.

(2) By simple must transfer from tank to tank, and from bottom to top of the tanks, splashing the must through air. This is much less effective than direct oxygen uptake. Experimentally, it is known that the total oxygen uptake needed by white must in order to oxidize the phenolic to precipitation is about (20 – 30) mg/L, so it will be necessary to do it several times at, say, one hour intervals to supply enough oxygen. If done at lower temperatures, e.g., (15 – 20) °C, the oxygen uptake will be higher but the oxidation rate will be lower.

Some further instructions are:

(3) It is recommended not to use bentonite at this stage (to clarify the must), because it reduces the enzymatic activity and the rate of oxidation. The process of such oxidation takes about (4 – 6) hours, where after, the must can be racked off its sediments. Now SO_2 can be added (with or without bentonite), followed by yeast inoculation to start the fermentation. When fermentation is over, the result is light-colored wine which is stable towards further phenolic oxidation. This means that the wine color will stay light, and bitterness will not develop on aging. But what about the aroma and wine quality? In principle, other constituents in the must might be oxidized as well during this process, but in view of the fact that most wine flavor components are bound to sugar or other molecules in the grape berries, and are released only during and after fermentation, the risk looks minimal. In regard to quality evaluation, the criteria are whether the wine is better than white wine made without this *macro-oxidation*, or whether there is any difference at all.

Many tastings have been done to find out what, if any, changes resulted from pre-fermentation oxidation, with controversial results, depending on the variety, and other techniques that are used in making white wines, when reference comparison was made. For example, if whole bunch press is gently done so that the amount of phenols is small, there might be no difference between the reference wine and the one with pre-oxidized must. Another possibility is that the amount of oxygen was not sufficient to precipitate the whole amount of phenolics. As for general wine quality, some wines were found better, some were worse, and with others there was no difference.

Acetaldehyde

It is well known that aldehydes in general, and acetaldehyde in particular, are formed during wine aging as a result of chemical oxidation of the wine's alcohols. Ethanol is stable towards oxidation when in contact with air oxygen at room temperatures (as are other alcohols), whereas in wine it is oxidized to acetaldehyde in a short time after exposure. Therefore, the oxidation mechanism must occur by an indirect reaction of the alcohol with oxygen through an oxidizing agent which was postulated to be *hydrogen-peroxide*. This was later confirmed by many studies:

(1)

(8) (2) CH_3-CH_2-OH + H_2O_2 $\longrightarrow$ CH_3-CHO + $2H_2O$

Step (1) represents any *ortho-dihydroxy-phenolic* in wine solution undergoing oxidation to *ortho-quinone* and *hydrogen-peroxide*, as was discussed above in this section. The hydrogen-peroxide formed is very reactive to oxidize any oxidizable component. Although ethanol is less reactive than a phenolic, its very high concentration in wine solution makes it a significant competitor in this case when a strong oxidant such as hydrogen peroxide is present.

In a work on acetaldehyde formation in a model wine solution it was shown that the maximum acetaldehyde that can be produced is limited by the oxidizable phenolic content, based on a molar ratio. That means that the ratio of acetaldehyde formed to phenolic content (in molar ratio) cannot exceed one (one molecule of acetaldehyde per one molecule of oxidizable phenolic), as is expected from the above coupled oxidation reactions (1) and (2). This is because for each oxidized molecule of phenolic, one molecule of hydrogen-hydroxide is also formed and reacts with one molecule of ethanol. It was shown in this work that the presence of ortho-dihydroxy-phenols is necessary for the production of acetaldehyde. The known phenomenon of "bottle sickness" that occurs after wine bottling is caused by acetaldehyde formation in the bottle, as a result of excessive exposure of the wine to air during bottling. The acetaldehyde "sickness" disappears after a while due to reactions with other components in the wine solution (sulfur-dioxide and phenols).

If enough free sulfur-dioxide is present in the wine, it can reduce the hydrogen-peroxide to water, and be oxidized to sulfate:

$$(9) \qquad HSO_3^- + H_2O_2 \Longleftrightarrow HO\text{-}OSO_2^- + H_2O$$
$$\downarrow$$
$$SO_4^= + H^+$$

This reduction-oxidation reaction is catalyzed by iron ions (Fe^{2+}/Fe^{3+}) and copper ions (Cu^+/Cu^{2+}) found in wine in concentration range of (3 – 15) mg/L and (0.1 – 3.5) mg/L respectively. The SO_2 acts as a scavenger to interrupt the oxidative activity of hydrogen-peroxide, and this is one way in which it acts against chemical oxidation.

2. Micro-oxygenation

Because of the essential contribution of oxygen to the development of wine bouquet and aging, a significant function in wine processing is fulfilled by wooden barrels. (For the full information on oak barrels, see chapter V, sections A and B.) Here we will mention briefly the two purposes achieved by aging wine in barrels. First, to enable slow exposure of the wine to air in order to maintain slow oxidation reactions of wine components, mainly its phenolic compounds. This is achieved by frequent topping of the lost wine which has evaporated through the wood. The small amount of oxygen that enters into the wine through the open bung and by the oxygen-saturated topped wine is enough to carry on the slow aging process discussed earlier.

The second reason for using barrels is to introduce some flavor from the wood that is compatible with the flavors of the wine. It turns out that in the winery economy the cost of barrels is of great significance. Besides the initial very high cost of barrels, their use is temporary and limited to a few years. The treatment and maintenance of the barrels also cost much more than that for stainless steel tanks. On top of that, barrels are more difficult to maintain clean than stainless steel, leading to microbial hazards such as *brettanomyces*, *acetobacter* and other microbial risks. Once knowing the total cost of using barrels, and understanding the function of barrels, the next question is why one should not achieve the same results in a stainless steel tank, with oak wood pieces (chips, staves, segments, or any size of wood) immersed in the wine, and by slowly infusing oxygen into the tank. By doing so the cost of producing wine can be significantly reduced.

On the basis of this idea, and on the exact monitoring and controlling of the oxygen intake, what is called *micro-oxygenation* emerged. Micro-oxygenation (MO) is a technical solution for these issues, mainly in the bulk wine market of medium to low prices. It started in France during the mid-1980s and spread all over the world.

Technically, it is done by releasing food-grade oxygen gas from a high pressure cylinder tank with a pressure control valve through plastic tubes to a micro-porous diffuser with channels size in the $(1-10)\,\mu$ range, made of ceramic or stainless-steel. Bubble size is set by the oxygen pressure and the diffuser pore size: the smaller the size, the smaller the bubbles. The dose of oxygen flow into the wine in units of mg/L/month is controlled by a special gas doser and by the pressure valve. The diffuser (typically a tiny cylinder with dimensions of a few mm in diameter and $(10-20)$ mm length), is placed at the bottom of the wine vessel. Small bubbles coming up from the diffuser against the hydrostatic pressure of the height of the wine above are quickly dissolved in the wine. The MO is done in stainless steel tanks for a certain period, after which the wine is ready for bottling as soon as all needed cellar operations are finished. In certain cases, after MO is done, the wine is transferred to continue barrel aging. The rationale of this is questionable. If the process is done to save money and time, then what is the purpose of using oak barrels anyway? Maybe the answer is to give the wine some extra boost of oxygen before barrel storage, in order to accelerate the aging. The principles of MO are listed below:

(1) There are two phases in the operation: The first one right after fermentation and until the end of MLF, which takes a few weeks, and the second phase after MLF is over, for a certain time of one to six months. Thereafter, the wine can be bottled or be aged in barrels if this is the choice.

(2) The use of SO_2 should be avoided during the first phase because it interferes in the oxidation process. One of the purposes of introducing SO_2 into wine is that it reacts with the acetaldehyde formed during oxidation and blocks it from further reactions. The formation of acetaldehyde during the oxidation through hydrogen-peroxide is important, because acetaldehyde induces binding between phenolic molecules to polymerize. SO_2 reacts with acetaldehyde to form a quite stable compound, thereby interfering in the polymerization process. Another reason for avoiding the use of SO_2 is not to interfere in the MLF. Only after the first phase is

over can SO_2 be added to support antimicrobial protection.

(3) How much oxygen is needed for oxidation?

Some data and small calculations can give us a very rough idea about it:

According to the basic oxidation reaction, one molecule of dihydroxy-phenolic reacts with one molecule of oxygen to create one molecule of quinone plus one molecule of *hydrogen-peroxide* (equation 8 in this section).

These two agents can initiate the oxidation process separately, which means that one oxygen molecule can initiate two different oxidation chains. It means that one molecule of dihydroxy-phenolic is basically oxidized by one molecule of oxygen. The second step is based on phenolic analysis of red wine, where we can take a typical total phenolic concentration in red wine as $(1 – 2)$ g/L. Let us take a middle number of 1.5 g/L. Red wines contain about 80% of the phenolic as flavonoids, so we are left with 1.2g/L to make the estimation. Out of the many flavonoids, we can use catechin as a representative molecule with molecular weight of M = 290. So, roughly each liter of red wine contains about 1200 mg of potential oxidative catechin, which is 1200 mg/290 ~ 4 mili-mole of catechin. To oxidize that amount of catechin it will require 4 mili-mole of oxygen: 4x32 ~ 120 mg. This means that **one liter of typical red wine will need approximately 120 mg of oxygen** to be significantly oxidized.

It was also observed many years ago that red wines can tolerate up to 180 mg/L of oxygen before showing over-oxidization character, with a recommendation of about 60 mg/L of oxygen for improved quality. If done the traditional way of oxygen saturation by just racking barrels and pumping, about 10 cycles of operation are needed; each one can saturate the wine with $(5 – 7)$ mg/L of oxygen. This gives us some idea in what range of total oxygen uptake we are dealing. Wines held only in big tanks have no chance to gain that amount of oxygen, and this is the point of using MO.

The common oxygen dosage rate in phase one after fermentation and before MLF is about $(20 – 60)$ mg/L/month. As stated earlier, the duration of MO at this phase is a few weeks. After MLF is over, sulfur-dioxide can be added and phase two of MO can be started at a lower dose rate of $(2 -10)$ mg/L/month for one to six months. Taking a typical MO use of 35 mg/L/month at phase one for one month, and then at phase two 4 mg/L/month for, say, six months, a total of about 60 mg/L of oxygen will be introduced.

In the second phase at a slow oxygen flow rate, it is recommended to monitor wine taste every week, in order to decide when to stop the MO. For comparison, when racking barrels or topping, there is no control over how much oxygen is dissolved in the wine. But from experience, **on regular barrel use which includes topping and racking, about (50 – 70) mg/L/year of O_2 is dissolved in the wine**. By using MO, the optimum rate of oxygen flow should be such that it will never accumulate in the wine but will be in constant balance with the oxidation reaction rate. Measurements show that dissolved oxygen concentration in wine during the MO process is in the (0.01 – 0.1) mg/L range.

(**4**) Some technical notes:

a. The temperature of wine while using MO is best in the (15 – 20) $^{\circ}$C range.

b. The commercial diffusers can supply oxygen from 0.2 to 150 mg/L/month.

c. Because some flow meters are in cc/L/month, the conversion is simple:
The volume of one mole of gas (any gas) measured at standard conditions, namely, atmospheric pressure at 0°C, is 22.4L (general law for any mole of gas).
The molecular weight of oxygen is 32. One mole of oxygen is 32 gram. Therefore, one ml of oxygen will weigh 32,000 mg/22,400 cc = 1.43 mg/cc. At room temperature (25°C) the volume will expand, and it will weigh 273/298 of its weight, namely, 1.3 mg/cc. So, when using MO with various meters, it worthwhile to remember this factor: **1ml = 1.3 mg**.

(**5**) Summarizing the advantages of MO:

a. It enhances red wine's color and stabilizes it.

b. Another advantage of MO is that it oxidizes small thiols such as methyl-sulfide to dimethyl-sulfide, and reduces some undesirable odor:

$$2CH_3\text{- }SH + 1/2O_2 \quad \Longleftrightarrow \quad CH_3\text{- S - S- }CH_3 + H_2O$$

This is a reversible reaction, and its direction depends on the oxidation-reduction potential in the wine. It happens regularly when exposed to air during wine treatment, and then after a period of time in the bottle, when the potential becomes reductive, the bad smell might come back. MO can be used in certain cases to correct problematic wines with non-specified off-odor.

c. One additional advantage of MO is that it saves aging time in the winery, and wines can be bottled and be released a short time after MO.

d. Although the practice is quite rare, some wineries use MO for white wines, mainly Chardonnay, to enhance fruitiness and aging bouquet, and prefer to do it *sur-lie* (in stainless steel tanks or barrels). In such a case the rate of oxygen flow is very low (about 1-2 mg/L/month) for about one month. In this connection, it should be remembered that pre-fermentation air exposure of white must (macro-oxidation) is practiced sometimes in white wine, and the reader is referred to the discussion earlier in this section.

e. And last, the main advantage of micro-oxygenation technique is its ability to control the total amount and rate of oxygen exposure.

One study which invested a huge effort to going deeply into the chemical changes happening during micro-oxygenation, compared a group of three **no micro-oxygenation** treatments to a group of three **micro-oxygenation** ones in Cabernet Sauvignon wine. The first non-oxygenated group was: one control, one with oak staves added, and one with oak segments (chips) added. The group of micro-oxygenated wines had the same protocol (one control, one stave added and one oak segments added). The oak pieces (staves and segments) were toasted and were added at 3 gr/L of wood. The experiments were done in an industrial scale of 10,000L to 40,000L each. Micro-oxygenation rate was 10 mg/L/month for one month and 5 mg/L/month for another 4 months. SO_2 was added three times during the experiments, 30 ppm each time. Analysis was made for almost any compound that could be changed during the experiment, in two different laboratories. Two parameters were studied in the treatments: one was the difference between micro-oxygenation treatment and without it, and the second was the difference between the uses of staves vs. oak segments. And of course, the mutual effects between the two parameters were also studied.

By grouping the individual results of each parameter and averaging them, some comparison could be made about the effect of each parameter under study. It seems surprising that although there were some differences in the concentration of the main components between the micro-oxygenation group and the one without it, these changes were very minor (!).

For example (concentrations of components are in mg/L, the numbers taken from the figures in the study), see the chart at the top of the next page. What can be seen here? Based on these numbers, which were carefully measured, the effect of MO is somewhat disappointing. All changes in the composition of the main components are quite minor. The main non-flavonoid representative *caftaric-acid* was reduced by 1mg/L (10%), and the main flavonoid *epicatechin* by 3 mg/L. Polymeric anthocyanins were increased by 7%, and the monomeric ones decreased by the same value. Total absorbance spectra reflect only a small change in color intensity. Perhaps the

211

Component	no-MO	MO	% change
polymeric phenols	650	720	10
polymeric anthocyanin	47.5	51	7
monomeric anthocyanin	51	47	8
caftaric-acid	12	11	8
epicatechin	23	20	15
total absorbance (520nm)	51	56	

measured changes by the micro-oxygenation treatment were so low because of the use of SO_2 during the experiments, right from beginning. And if so, it shows the efficiency of sulfur-dioxide. Also, it is probable that the total of 30 mg/L of oxygen that was introduced in this MO study was too low, as was the very low intake rate. Until more studies have been done, the real contribution of MO and its benefits to winemaking will remain unclear.

To sum up the issue of MO, at the high end of wine quality, it cannot replace the "old" tradition of aging wine in barrels, where the wine is more complex and has better aging potential. In other words, it does not "make" better wines. But for the bulk market, which is the majority of the wine trade according to many studies in the literature, MO has the advantages of producing better balanced, softer, and cleaner wines. But, by operating MO, some of the specific aroma components which express the varietal and regional characteristics may be "gone with the wind of oxidation."

And a personal opinion on this matter: I did not carry out specific research of any kind in wines made by this method, but I happened to taste a series of such red wines (with two very knowledgeable colleagues of mine) and we came to the following conclusions: One, all of the wines were very well made. They were fruity, balanced, and enjoyable. Second, although they were from all over the globe, they were the same, as if there were no local characteristic and even almost no varietal one. Globalization at its best. The rest is a matter of opinion, not a professional judgment.

C. Sulfur-Dioxide

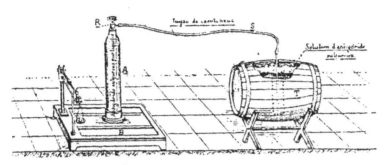

PRÉPARATION D'UNE SOLUTION D'ANHYDRIDE SULFUREUX.

1. Sulfur-dioxide as a Food Products Preservative

Sulfur dioxide is a colorless gas, with a very strong, suffocating and irritating odor. Some of its physical properties are:

molecular weight: 46.07;

liquid form: under one atmosphere pressure at -10°C

under 3.25 atmosphere pressure at 20°C.

Sulfur dioxide has been used as a preservative agent in the food industry for a very long time, probably since the Roman era. In wine, sulfur-dioxide has been used from the Middle Ages up to the present. It seems almost impossible to produce wine without the addition of sulfur dioxide, for reasons which we will discuss in this section.

Molecular sulfur-dioxide (SO_2), bisulfite (HSO_3^-), and sulfite ($SO_3^=$) are multi-action agents for the following functions:

Antioxidant agent – in various systems, by reducing oxidation products such as hydrogen-peroxide, quinone, acetaldehyde and others. In most cases it blocks oxidation reactions that cause damage to the food product to continue. In such reactions the sulfite enters as a derivate to the organic compounds, and its ($SO_3^=$) form is oxidized to sulfate ($SO_4^=$).

Inhibition of microorganism growth – in food products, mainly bacteria, and in higher concentrations also molds and various species of yeasts. The exact mechanism of inhibition is not yet known, but the molecular form of sulfur-dioxide in water (as SO_2) is the active form.

SO₂ chemistry in aqueous solution

The equilibrium of sulfur-dioxide in water is carried on in two steps:

$$SO_2 + H_2O \Longleftrightarrow H^+ + HSO_3^- \quad ; \quad HSO_3^- \Longleftrightarrow H^+ + SO_3^=$$

The first one is an equilibrium between molecular SO_2 and bisulfite ion (HSO_3^-), and the second one is between the bisulfite ion and sulfite ($SO_3^=$).

At lower pH range where the first equilibrium is dominant, the dissociation constant K_a of the reaction is defined as:

$$K_a = [H^+] \times [HSO_3^-] / [SO_2]$$

Where: $[SO_2]$; $[HSO_3^-]$; $[H^+]$ are the molecular SO_2; bisulfite ion; and hydrogen ion concentrations respectively.

And because the value pK_a is formally defined as: $pK_a = - \log K_a$;

and also $pH = - \log[H^+]$; $\Longrightarrow pK_a = -\log K_a = -\log[H^+] \times [HSO_3^-] / [SO_2]$

$= -\log[H^+] - \log\{[HSO_3^-] / [SO_2]\} = pH - \log\{[HSO_3^-] / [SO_2]\}$; therefore:

$$(11) \qquad \boxed{\log \{[HSO_3^-] / [SO_2]\} = pH - pK_a}$$

which means that the ratio of the *bisulfite form* (HSO_3^-) to the *molecular form* $[SO_2]$ of sulfur-dioxide, is pH dependent (!) and is equal to the pH of the solution minus the constant number pK_a. It is obvious that this concentration ratio can be calculated, once the value of pK_a is known. Also as a result of this equation when (by chance) the pH of the solution is equal to the pK_a, then the ratio $\log[x] / [y] = 0$; $\Rightarrow [x] = [y]$; or the two forms bisulfite $[HSO_3^-]$ and $[SO_2]$ are equal. This means that the pK_a can be a marker, at what pH there are equal concentrations on both sides of the equilibrium equations.

In the second of the above equilibrium between bisulfite and sulfite anions, at higher pH range where the second equilibrium is dominant, the dissociation constant K'_a is by definition: $K'_a = [H^+] \times [SO_3^=] / [HSO_3^-]$

and $pK'_a = -\log[H^+] - \log\{[SO_3^=] / [HSO_3^-]\} = pH - \log\{[SO_3^=] / [HSO_3^-]\} = pH + \log\{[HSO_3^-] / [SO_3^=]\}$;   therefore:

$$(12) \qquad \boxed{\log \{[HSO_3^-] / [SO_3^=]\} = pK'_a - pH}$$

This equation describes the concentration ratio of *bisulfite form* (HSO_3^-) to the *sulfite form* ($SO_3^=$) of sulfur-dioxide in the equilibrium equations.

For the above equations the two equilibrium constants are (at 25°C):

$$pK_a = 1.81 \; ; \; pK'_a = 7.2$$

Note the significant difference in the values of pK_a and pK'_a—they are the log figures of the dissociation constants of sulfur-dioxide in water solution, and therefore they represent roughly 10^5 equilibrium ratio between the first dissociation to the second one (7.2 - 1.81 ~ 5).

The **molecular SO_2** form acts as an **antimicrobial agent**. In order to determine its actual concentration in solution (out of the total concentrations of all three forms), let us state the definition of *"free"* sulfur-dioxide. This term includes the *concentrations* of all sulfur-dioxide forms which are in equilibrium as shown above, and are not bound to any compound presents in the solution.

(13) $\boxed{\text{Free } SO_2 \text{ is the sum of } [SO_2] + [HSO_3^-] + [SO_3^=]}$

Close examination of the system in two pH ranges will reveal that:

a. In wine where the pH range is (**2.9 – 4.0**), the concentration of $[SO_3^=]$ is so small compared with the other forms (because of the pK'_a value = 7.2 is very far from wine pH), that for all practical *analytical* purposes, free sulfur-dioxide can be defined as: $[SO_2]_{free} = [SO_2] + [HSO_3^-]$, which is the actual value that is analytically measured as free SO_2 by titration. Inserting the last expression of bisulfite into the formulation (11):

$\log \{[HSO_3^-] / [SO_2]\} = pH - pK_a$ for [bisulfite]/ $[SO_2]$ ratio ;

$\text{Log} \{((([SO_2]_{free} - [SO_2]) / [SO_2]\} = pH - pK_a = pH - 1.81 ==>$

Antilog both sides: $([SO_2]_{free} - [SO_2]) / [SO_2] = 10^{[pH - 1.81]}$

$[SO_2]_{free} / [SO_2] - 1 = 10^{[pH - 1.81]} ===>$

$[SO_2]_{free} / [SO_2] = 1 + 10^{[pH - 1.81]}$ and therefore:

(14) $\boxed{[SO_2] = [SO_2]_{free} / (1 + 10^{[pH - 1.81]})}$

b. At higher pH values (around 7.0) where the second equilibrium is dominant, molecular form $[SO_2]$ is so small (again because the pK_a =1.81 is far away from the actual pH) that the free sulfur-dioxide can be written as:

$[SO_2]_{free} = [HSO_3^-] + [SO_3^=]$

And inserting it into formulation (12) for the [bisulfite-ion]/[sulfite-ion] ratio:

$\log \{[HSO_3^-] / [SO_3^=]\} = pK'_a - pH \implies$

$\implies \log \{([SO_2]_{free} - [SO_3^=]) / [SO_3^=]\} = pK'_a - pH = 7.2 - pH$;

(converting again the log to its antilog):

$([SO_2]_{free} - [SO_3^=]) / [SO_3^=] = 10^{[7.2 - pH]}$

$[SO_2]_{free} / [SO_3^=] - 1 = 10^{[7.2 - pH]} \implies [SO_2]_{free} / [SO_3^=] = 1 + 10^{[7.2 - pH]}$

therefore:

$$(15) \qquad \boxed{[SO_3^=] = [SO_2]_{free} / (1 + 10^{[7.2 - pH]})}$$

These last two formulations (14, 15), enable us to calculate the molecular SO_2 concentration and the sulfite-ion ($SO_3^=$) at any pH of interest, by knowing the free SO_2 concentration (by titration). And vice versa, we can estimate the free SO_2 needed to maintain a certain value of molecular SO_2 or $SO_3^=$ at a given pH. For example, to maintain 0.8 ppm SO_2 at pH = 3.5, the free SO_2 should be:

$[SO_2]_{free} = (1 + 10^{[pH - 1.81]}) \times [SO_2] = (1 + 10^{[3.5 - 1.81]}) \times [0.8] =$

$= (1 + 10^{1.69}) \times [0.8] = (1 + 49) \times 0.8 = \textbf{40 ppm}$.

At pH = 3.8 (0.3 pH units higher than the last example) the free $[SO_2]$ needed for the same concentration of molecular SO_2 (0.8 ppm) is about **80 ppm**.

The distribution of the three forms of sulfur-dioxide in various pH solutions can be calculated on the basis of equations (13) ----> (15) and are shown in the following figures:

Molecular SO₂
Molecular sulfur-dioxide concentration
at wine pH range

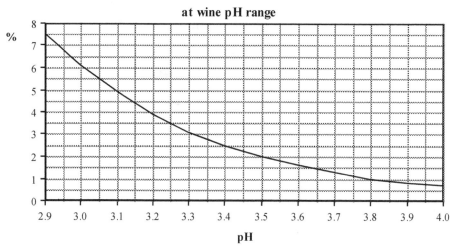

Bi-sulfite HSO₃⁻
Bisulfite ion concentration

at wine pH range

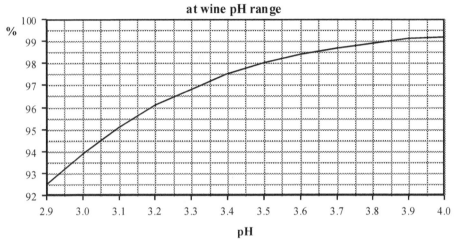

Sulfite SO₃⁼
Sulfite ion concentration

at wine pH

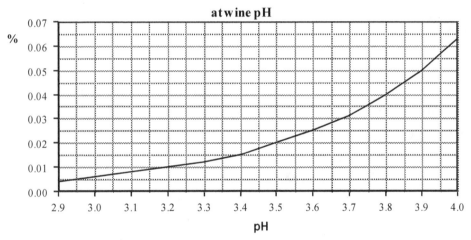

Most of the sulfur-dioxide $(90-99)$ % in this pH range exists as bisulfite ion (HSO_3^-) while the molecular form (SO_2) is at about $(1-7)$ %. The sulfite ion $(SO_3^=)$ exists only at a very low concentration range of $(0.01 - 0.1)$ % range.

For the general interest, the full picture of the distribution of the three forms of sulfur-dioxide, namely, molecular (SO_2), bisulfite (HSO_3^-) and sulfite $(SO_3^=)$, in a wider pH range can be calculated by using the above formulae, and is shown in the following figure:

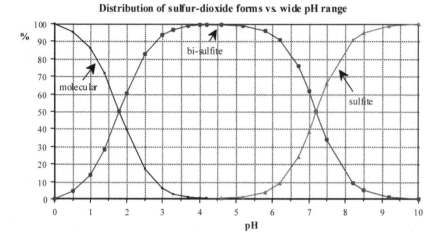

Distribution of sulfur-dioxide forms vs. wide pH range

(Note the equal concentrations of *molecular* and *bisulfite* forms at pH = 1.81; also *bisulfite* and *sulfite* forms at pH = 7.2 as expected by their appropriate pK_a. Pay attention to the maximum percentage of bisulfite ion at pH range (4 – 5) which is the pH range of transformation between bi-sulfite ($\sim 100\%$) and sulfite ion ($\sim 0\%$).

SO_2 Toxicity

Sulfur-dioxide is a chemical agent which is well situated in modern life. We are exposed to it through many food products as food additives and by our atmosphere. The data on SO_2 in foods is very diverse in different sources, but typical contents in some foodstuffs are: dried vegetables (100 – 500) ppm, wines (50 – 150) ppm, salad bar (400 – 600) ppm, and dried fruits (1,000 – 2,000) ppm !!! Some average contents in particular examples are given in the figure at the top of the next page.

The evolution of sulfur-dioxide into our atmosphere from cars and industrial plants is estimated at 150 million tons annually. The standards in the U.S. dictate that the average atmospheric concentration may not exceed 0.03 ppm. The exposure to atmospheric SO_2 by respiration contributes an average intake of about 1mg/day. A joint committee of FAO and WHO established (in 1974) an *acceptable* daily intake (ADI) of sulfites (as SO_2) to be 0.7 mg/kg/day, which is about 45 mg/day for a 140-pound person.

Sulfite is metabolized in mammals mainly by molybdenum-protein enzyme (sulfite-oxidase), which oxidizes the sulfite to sulfate. The sulfate is then excreted

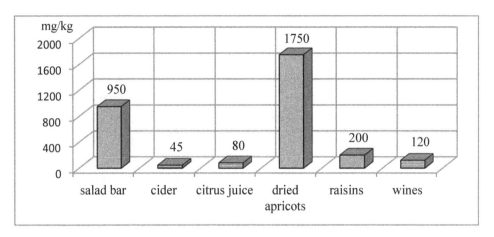

Sulfites content in some typical food products (as total SO_2)

from the body in the urine. The sulfite-oxidase enzyme has a most important function in metabolizing sulfites, which are metabolic products of the sulfur-containing amino acids cysteine and methionine. The enzyme prevents sulfite poisoning of the body. Another (minor) pathway of sulfite metabolism is through thiosulfate formation in cases of sulfite-oxidase deficiency. The mechanism of *endogenous* (inside body sources) sulfite metabolism is used also to metabolize the *exogenous* (outside-sourced) sulfites from food and environment uptake. The estimated daily consumption of sulfite from food products in Europe (including 300 ml of wine) is about 60 mg. The estimated human production and excretion (via urine) of *endogenous* sulfite is about 2,000 mg/day. These figures put the exogenous exposure to sulfites into some perspective (!)

The toxicity of sulfite to humans was studied in experiments which were carried on *below acute* doses (of course), up to 190 mg/kg (as Na_2SO_3 which was administrated intravenously). Not enough information is known about *bound* sulfites in the body, but in the last report, up to 1,200 mg/kg of SO_2 was orally administrated to rabbits as acetaldehyde hydroxyl-sulfonate. Another report states that subchronic administration of sulfites has no effect (in humans).

Chronic toxicity of sulfites has been studied in many works. In a long-term study in rats (up to three years, which followed multiple generations), no effect on growth, fertility, or weight of newborns was observed. Also *no pathological* signs were observed. In another study on rats and pigs, the no-effect level was observed at an uptake of 72 mg SO_2/kg/day for two years. The acceptable daily intake (ADI) mentioned above was established on the basis of this study by applying x100 times safety factor. *No carcinogenicity* was found to occur due to sulfites, nor was a *mutagenic* effect noticed in other studies.

Another aspect of the SO_2 influence on human health is its *allergenic* effect. Many papers have been published in recent years, reporting the induced asthma caused by sulfite administration. Some safety conclusions could be deduced. There is a sensitivity of a certain portion of the population with asthma to other agents, as well as to SO_2. Also, a distinction in population sensitivity has to be made between ingestion of sulfites (in food or capsules of the agent) and acidic solution administration of sulfite. The last one was clearly shown to be a response to inhaled SO_2 and not to ingested SO_2. Also the doses to which sensitive people have to be exposed, in order to show symptoms of asthmatic reactions, are not so minute, in the order of $(25 - 100)$ mg of $K_2S_2O_5$. To summarize the issue, based on the huge studies that were taken on this matter, *an expert panel on food safety, the World Health Organization (WHO), reviewed the issue of sulfite use in 1975, and allowed it to remain in the category of "Generally Recognized as Safe" (GRAS) list of food additives.*

(If the reader would like to see all the literature references related to the studies mentioned above, he or she is referred to my *Concepts in Wine Chemistry* book, chapter VII, section A).

2. Sulfur-Dioxide Uses in Wine
The Effect of SO_2 on Microbial Activity

In this section we will discuss the specific aspects of sulfur-dioxide uses in wine production. Many yeast species are active in wine fermentation, such as *Saccharomyces-cerevisiae*, *Saccharomyces-bayanus*, *Kloeckera-apiculata* and *Candida-stellata*. At the beginning of "natural" fermentation, the so-called "wild yeast" species are mostly active. Later, the more alcohol-tolerant species *Saccharomyces-cerevisiae* and *Saccharomyces-bayanus* take over to complete the fermentation. In "cultured yeast" inoculation (based on the last two species) the specific yeast strain used for inoculation is practically the dominant one during fermentation, for two possible reasons: it is the most abundant in the culture, and because of the SO_2 addition to the must. In white must with up to 100 ppm of SO_2 added before fermentation, there is almost no practical effect on the *"cultured yeast"* growth, whereas at 50 to 100 ppm SO_2 addition, there is a delay of a few days in the growth of the *"wild yeast."* (Addition of 150 ppm of SO_2 is needed to completely suppress the growth of the *"wild"* species.) In red must, on the other hand, with addition of 100 ppm of SO_2 no difference was observed between *inoculated* and *un-inoculated* fermentations. This is due to the presence of anthocyanins in red wines, which interact with the SO_2, reducing its inhibitory effect on yeast growth. In view of these findings and the practice of SO_2 addition during fermentation at about 20 to 60 ppm, and the use of "cultured yeast" strains, it appears that the control against "wild yeasts" is not necessary and not effective. The effect of inhibition or delaying inoculated fermentation at these SO_2 addition levels is insignificant.

The effect of SO_2 on the growth of *Saccharomyces-cerevisiae* yeast species was studied by experiments with radioactive SO_2. It was found that the molecular SO_2 is the sulfite form that is transported into the cells, a process that is completed a few minutes after application. Also it was found that the survival of the yeast cells after SO_2 application does not depend on the total SO_2 but on the molecular form. The molecular SO_2 concentration range depends, of course, on the pH, and for example (at pH = 3.19) a zero survival of the yeast cells after 5 minutes of application was achieved at 20 mili-mol of added SO_2 (about 1300 ppm) which is approximately a concentration of 50 (!) ppm of *molecular* SO_2 at this pH (about 4% of the total sulfur-dioxide in solution; see the first figure in this chapter above).

Practicing lower levels of added SO_2 in winemaking, on inoculation with *Saccharomyces-cerevisiae* yeast, may result in a small delay in starting the fermentation and also in completion time, mainly at addition levels above 50 ppm, as can be seen in the following figure:

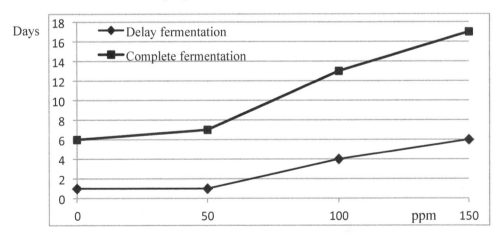

The effect of adding sulfur-dioxide to must, on delay of starting
fermentation and time of completion in S. cerevisiae at 21°C.

It seems that up to 50 ppm there is practically no effect on delaying and completing of the fermentation. Between 50 to 100 ppm, which is in the very upper range of sulfite addition practice before fermentation, there is some delay (up to 4 days at 100 ppm addition), and an additional week is required to complete it.

On the other hand, there are some yeast species which are very tolerant towards high SO_2 concentrations, causing wine spoilage (e.g., *Saccharomyces-bailii*), and others that can maintain a normal fermentation rate with up to 300 ppm of added sulfur-dioxide. In secondary fermentation of sparkling wine, the free SO_2 level is very important with regard to problems that may arise in maintaining the fermentation. Levels of (25 – 30) ppm of free SO_2 are high enough to reduce very significantly the viable yeast population. The results may be a long delay in starting the fermentation, very slow fermentation, and eventually stuck fermentation with low carbon-dioxide pressure. A safe free SO_2 level in secondary sparkling wine fermentation is 10 ppm.

The preservation of grape juice from fermenting is of increasing interest in cases where the juice is stored for later fermentation, or as a sweet additive to white wines. Two methods can be applied to preserve the juice: very high concentrations (1200 – 2000) ppm of SO_2, where the juice is stored at ambient temperatures, and lower SO_2 concentrations of (200 – 300) ppm where the juice is kept at low temperatures (0 – 5) °C. In the first case, when the juice is needed for use, it has to be de-sulfited by heat and by applying a nitrogen stream through it, leaving the juice with only about 100 ppm of SO_2. In the second case, fermentation can be delayed for a few months under these conditions. No mold or other spoilage bacteria (acetobacter, or ML bacteria) grow under such conditions. Malolactic bacteria, which play an important part in winemaking practice (encouraging or preventing), is inhibited by the presence of SO_2, free and bound. (The dependence of ML fermentation rate on the total SO_2 concentration and pH values, can be seen in the chapter III, section D).

Interactions of SO_2 with Wine Components

Only part of the total SO_2 concentration exists as *free* SO_2. The other part is bound to different components present in the must or wine. The amount of the added sulfur-dioxide that remains as free SO_2 in white *must* (Riesling) when it was added before fermentation, is demonstrated in the following figure:

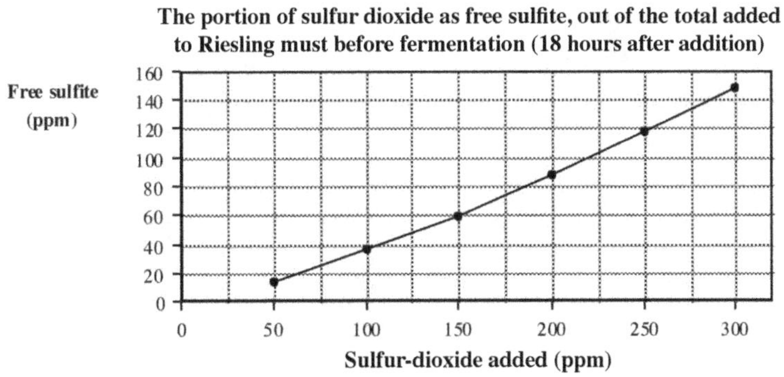

**The portion of sulfur dioxide as free sulfite, out of the total added
to Riesling must before fermentation (18 hours after addition)**

Only about 40% of the sulfur-dioxide added to the must is left as measured free SO_2. And this pattern is almost linear all the way up to very high additions. The rest is bound to the must components in a very short time.

To get some idea how much free SO_2 remains in a newly fermented white *wine* (Gewurztraminer) 5 days after addition of the same quantities of sulfur-dioxide, see the following figure:

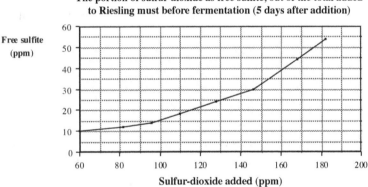

**The portion of sulfur dioxide as free sulfite, out of the total added
to Riesling must before fermentation (5 days after addition)**

(No addition of SO_2 was made before fermentation.)

Clearly, most of the added sulfur-dioxide, especially at lower concentrations, does not remain as free SO_2. At 60 ppm addition, only 10 ppm was left after five days where the measurement was done. Only at very high additions beyond the usual practice of vinification, free SO_2 remains above 40%. Hence most of the SO_2 is bound and not available for its expected usage. The major components in must and young wines which interact with SO_2 to bind it are carbonyl compounds, phenols, pigments, and some other specific components.

Interaction with Anthocyanin

Anthocyanins are positively charged in the central ring. Therefore they are susceptible to attack by a negative agent such as bi-sulfite (HSO_3^-):

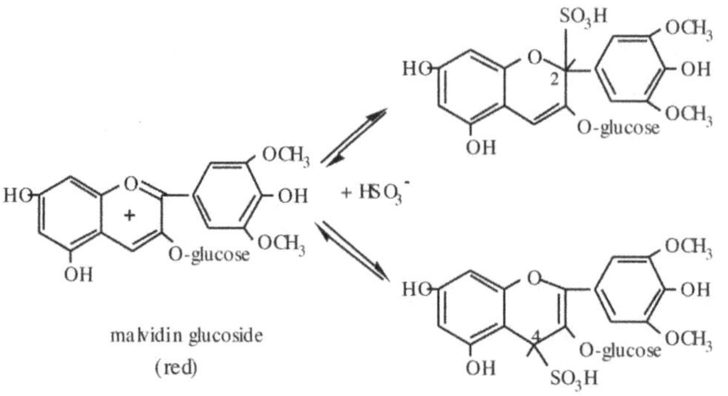

malvidin glucoside
(red)

$+ HSO_3^-$

malvidin glucoside hydroxysulfonic acid
(colorless)

By this interaction the double bond between carbons 3 and 4 is opened and the HSO_3 group takes place on either of these carbons, where position 4 is preferred (see figure). Therefore, co-polyphenols of anthocyanin and tannin, in which the polymeric bonding is at the 4 position, fail to react with sulfites. Note that the reaction is *reversible*, and under suitable conditions the bound sulfite (and pigment) may be released. The dissociation constant of the *anthocyanin-sulfite* equilibrium:

$$F^+ + HSO_3^- \Longleftrightarrow FHSO_3$$

is: $K_d = [F^+][HSO_3^-]/[FHSO_3]$; (where F^+ is the anthocyanin pigment ion and $FHSO_3$ is the sulfited-anthocyanin). This anthocyanin-sulfite constant was studied in wine pigments (cyanidin, malvidin, delphinidin, pelargonidin) and was found to be at about 4×10^{-5} in wine pH range (3.0 – 3.5) at 20 °C for all pigments. This small value of the dissociation constant means that the equilibrium is maintained mostly toward the bound side (right). A last note in this issue, is that the sulfited anthocyanin is colorless.

Methods of SO_2 Applications

As mentioned above, the antimicrobial activity of sulfur-dioxide to most wine-related micro-organisms is caused by the molecular form (SO_2). Its effective concentration range on microbial activity is about (0.5 – 0.8) ppm. The sulfur-dioxide forms are pH-dependent; therefore the concentrations of free sulfur-dioxide that are needed in order to maintain the above SO_2 values are shown in the following figure:

Free sulfur-dioxide concentration needed to maintain 0.8 ppm and 0.5 ppm molecular sulfur-dioxide at wine pH range

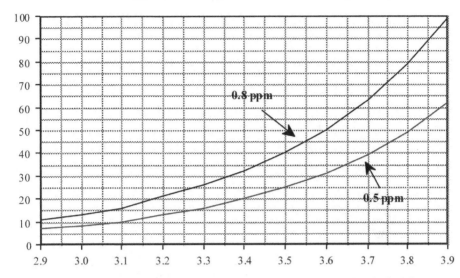

(This figure should be hung in front of the winemaker's desk!)

It is recommended therefore to store wines during cellar operations with free sulfur-dioxide concentration values that are within the area between the two lines. For example, for a wine with pH = 3.5 the recommended range of free sulfur-dioxide concentration at bottling should be (25 – 40) ppm (see figure above).

As mentioned before, when SO_2 is added to wine (or must), part of it reacts with some of the wine's components to become bound and therefore inactive as an antimicrobial and an antioxidant agent. The rate of binding depends on state and age of the wine. It is very rapid when added to must (requiring only a few minutes), and becomes slower as the wine gets older. The process of SO_2 becoming bound continues, with the result of a constant decreasing of free SO_2 in wine at any stage of its life.

Additions of SO_2 are made at different stages of vinification. At crushing, about (0 to 60) ppm is added to the must, depending on various considerations such as:

a. *Wine pH*. The higher the pH, the greater the sulfite addition needed to be effective (see figure above).

b. *Grape condition*. The moldier the grapes, the higher is the sulfite content that becomes bound by the mold substances. In such cases, the amount of SO_2 added should be twice that for healthy grapes. Also, higher concentration is needed to protect the must against enzymatic oxidation by *laccase* in high-infected grapes.

c. *Malolactic fermentation*. Because ML-bacteria are sensitive to SO_2, when malolactic fermentation is desired during fermentation, SO_2 addition should be minimal or not made at all.

d. When *intentional oxidation* and subsequent precipitation of phenolic components in white wines is desired, no addition of SO_2 is done at crushing. After fermentation is over, a very small (if any) concentration of free SO_2 is found in the wine, regardless of the initial addition, and most of the bound SO_2 (unless very high amount was added) is lost by oxidation to sulfate and by precipitation with the pomace particles and yeast cells. At first racking and afterwards during the whole process of winemaking, additions of SO_2 should be made according to the above figure to maintain $(0.5 - 0.8)$ ppm free SO_2. After fermentation, each addition is partially accumulated as *total* SO_2. The rest is lost (cannot be detected by SO_2 analysis) to become sulfate ($SO_4^=$). A few days after any addition, analysis of SO_2 (free and total) should be made. Care is recommended not to exceed a total SO_2 above $(150 - 200)$ ppm (although the legal limit in most countries is about 250 ppm).

It should also be mentioned that even without any SO_2 addition, some sulfite is formed during the alcoholic fermentation by the reduction of sulfate. The amount that is formed depends on the yeast strain, and it was found to be in the range of $(10 - 30)$ ppm.

The most practical usage of sulfur-dioxide addition is by:

(1) Solid salt of *potassium-meta-bisulfite* (or sodium salt) which maintains the following equilibrium in water:

$$S_2O_5^= + H_2O \iff 2HSO_3^-$$

The HSO_3^- ion is in equilibrium with all other forms of sulfur-dioxide, namely, SO_2 and $SO_3^=$ as shown at the beginning of this section. And therefore an equilibrium is set up that contains all three forms of sulfur-dioxide in accordance with the pH.

The molecular weight of potassium-meta-bisulfite ($K_2S_2O_5$) is (M = 222). It contains two molecules of SO_2 (with M = 64). Hence, 222 gram of the salt contains 128 gram SO_2 (57.6%). To add 100 mg/L of SO_2, 222/128 = 174 mg of the salt will be needed. A very useful short formula may help to calculate the amount of potassium-meta-bisulfite needed for a given volume of wine, in order to add the desired mg/L (ppm) of sulfur-dioxide:

(15) | **17.4** *gr of potassium-meta-bisulfite/HL* **==> 100 ppm**

(For example, to add 40 ppm of sulfur-dioxide to a tank containing 8000L (80HL), one should take 17.4x40x80/100 = 557 gram.)

(This very useful formula should also be hung on the winemaker's desk!)

(2) Liquid SO_2 (density 1.4 gr/cc at 15°C) from steel tanks which are held at (3 – 4) atmospheres pressure. When the gas is released into the wine, it is almost totally dissolved as sulfurous acid, and the dosage can be monitored by weight losses of the steel tank or by a flow dosimeter.

To sum up this section, the question whether to use SO_2 in wine and how much, if observed with unbiased eyes, is that no real proof has been found to justify its elimination in food in general, and in wine in particular, because of its very low concentration in this case. It appears that the war declared against it has reasons other than public health. Otherwise, the warning "containing sulfite" would be enforced on any food product containing it, with a long list of such products. Also, the long history of winemaking has shown that wine made *without* the use of SO_2 is lower in quality than wine made with careful and professional management of SO_2 additions. Such wines (without SO_2) have a short lifetime and are unable to age for many years.

D. Wine Spoilage

Being a food product, wine is very susceptible to various chemical and microbial spoilages. However, wine spoilage (in contrast to other food spoilage) can only cause faults in appearance, smell, and taste, rather than constituting a hazard to health. The damage caused by wine spoilage is therefore only an economic and image issue. The name of this section could also be *wine instability,* chemical and microbial (not including *protein* and *potassium-bitartrate* instabilities, which were discussed in the Cellar Operations chapter sections B and C). The term "instability" is used to refer to a process of undesirable compositional changes that occur in the bottled wine, which make its smell, taste, and appearance undesirable or even unacceptable. Recognizing potential risks, preventing problems and curing them if possible (once they happen) will be the subjects of this section. Only the most important and common faults will be mentioned here.

1. Chemical Faults

Metal Haze: This fault is quite rare these days. Although it still happens occasionally, it can be easily prevented. The metals of concern are iron, copper and to lesser extent aluminum. Although these metals are present in grape juice in the range of $(1-7)$ ppm of iron and $(0.1-0.4)$ ppm of copper, their concentrations are so low, that metal haze problems can happen only by introducing them into the wine during vinification, for example, from equipment and accessories made of materials containing these metals. Contact of the acidic wine with such materials (iron, brass, bronze, and aluminum) dissolves some of them into the wine. The best way to avoid this problem is not to use any equipment or gadget containing these metals. The materials of choice in the winery are stainless steel, plastic, wood, rubber, etc.

(1) Iron: Iron haze is formed by the ferric ion under oxidative wine conditions. In white wine the haze is composed of *ferric-phosphate* (called "white casse"). In red wine it is composed of *ferric-tannate*, which is an iron-tannin complex (called "blue casse"). The maximal iron concentration where iron haze may *not* be formed is $(5-10)$ mg/L. Above this level the potential risk is significant. To detect iron haze in wine, the cloudy tested wine sample should be acidified by adding 1/4 of its volume with 10% HCl. If the haze disappears, *metal* haze is of significant probability. Addition of a few drops of potassium-ferrocyanide (5%) into the acidified sample will indicate the presence of iron in the wine sample if a blue color appears.

(2) Copper: Copper haze is formed under reduced wine conditions, by interaction of the copper ion with proteins (in white wines). The maximum safe concentration of copper is about (0.5 – 1.0) mg/L. The protein content in wine also affects the potential haze formation. Besides equipment containing copper, another potential source of copper in wine is the addition of copper sulfate to cure a hydrogen-sulfide fault. Careful and exact quantities should be added in order not to create a potential copper haze in the treated wine. To detect copper haze in wine, the cloudy wine sample should be acidified by adding 1/4 of its volume with 10% HCl. If the haze disappears, it is most probably a *metal* haze. To verify it as copper haze, take a new sample of the cloudy wine and add a drop or two of (3 – 10) % H_2O_2. If the haze dissipates it is probably copper haze. To complete the detection, a few drops of potassium-ferrocyanide (5%) should be added. Formation of red color indicates the presence of copper.

Hydrogen-Sulfide: This is a very common fault in winemaking. The typical smell resembles rotten egg. Most volatile sulfur compounds have a distinctive smell, with very low thresholds, and being a part (fortunately a very small part) of wine composition, they contribute significantly to the off-flavor and nose of wine. Sulfur for hydrogen-sulfide production may come from three sources:

(1) Yeast protein degradation which releases hydrogen-sulfide.

(2) Residue of elemental sulfur dust left on the grape skin from anti-fungicide treatments in the vineyard.

(3) Reduction of inorganic sulfur compounds such as sulfate ($SO_4^=$) and SO_2. The concentration of hydrogen-sulfide found in wine after fermentation ranges from trace amounts to about 1000 mg/L, with a typical value of a few hundred mg/L. Its off-odor threshold in wine and other beverages is quite low (values of about 50 – 100 μg/L). In air the detection threshold is about 10 μg/L. More specifically, the common causes of hydrogen-sulfide formation are:

(1) *Protein catabolism by proteolitic enzymes:* Such protein degradation of sulfur-containing amino acids (cysteine, methionine) may release hydrogen-sulfide. The major factor causing production of hydrogen-sulfide is *nitrogen* demand for the yeast growth. When this demand is not fully supplied, the nitrogen is gained by protein degradation processes which also release hydrogen-sulfide as a byproduct from sulfur-containing amino acids.

In order to control and limit hydrogen-sulfide production via this route, certain measures can be taken. Because the main source for hydrogen-sulfide formation is protein degradation, a decrease in protein content might limit its formation. But at the same time, a sufficient nitrogen supply, as free amino acids or as ammonia, will reduce the proteolitic activity needed to supply that demand, and therefore will reduce hydrogen-sulfide production. An adequate supply of nitrogen by diammonium-phosphate (DAP) addition prevents almost completely the proteolitic process during fermentation, and as a result controls very effectively the production of hydrogen-sulfide. The following figure shows its effect:

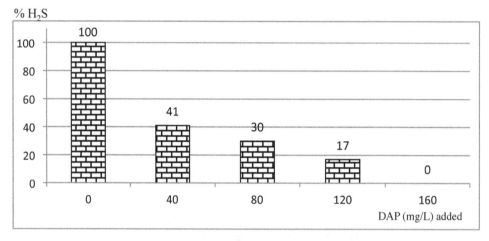

The effect of DAP addition to must before fermentation on formation of H_2S. The absolute value of 100% H_2S in this case is 422 mg/L.

DAP addition is also effective even when added into the must *after* the hydrogen-sulfide is being formed. A few hours after addition (while the fermentation is still active), the evolution of hydrogen-sulfide stops, and the bad smell disappears. The quantities of DAP that are needed to be effective are (as seen in the above figure) in the range of (100 – 200) mg/L.

As a consequence, preventative addition of (100 – 200) mg/L of DAP prior to fermentation is a regular procedure in many wineries. The addition of a small quantity of bentonite at the same time, to reduce the amount of protein, is also worthwhile.

(2) *Elemental sulfur:* Sulfur is used to dust the vines against *powdery-mildew* and other infections, and some of it remains on the grape skins. During fermentation and afterwards, the elemental sulfur (S^O) may be reduced to sulfide ($S^=$). Experiments with adding elemental sulfur dust to a fermenting medium at a sulfur concentration of up to 100 mg/L resulted in formation of up to 1200 mg/L of hydrogen-sulfide. The major factors that control hydrogen-sulfide production

from elemental sulfur are: the temperature (lower production at lower temperatures), elemental sulfur concentration (higher hydrogen-sulfide concentration at higher sulfur concentration, with linear dependence at the range of up to 100 mg/L sulfur concentration), and ethanol concentration (higher hydrogen-sulfide formation at higher ethanol concentration, probably by increasing sulfur solubility).

(3) *Inorganic sulfate and sulfite undergoing reduction to sulfide:* Sulfate ions $(SO_4^=)$ in the must, and sulfite added as sulfur-dioxide (SO_2), can be reduced enzymatically by the yeast cells to sulfides $(S^=)$. This reduction is done in order to supply a sulfur source for the production of sulfur-containing amino acids. When there is some shortage in the formation of these acid precursors, hydrogen-sulfide accumulates during fermentation.

(4) *Yeast strains:* One of the characteristics of yeast strains is their tendency to form hydrogen-sulfide, which is typical to any strain. Some are high producers and some are low producers. Most "wild yeasts" produce higher levels of hydrogen-sulfide. When choosing a yeast strain to carry on the fermentation, this characteristic may be of certain importance in a specific case. Usually the companies who produce yeast publish the characteristic details of each strain they produce, including their tendencies to form hydrogen-sulfide.

(5) *Metal ions:* When they are in excess, mainly copper (but iron, too) stimulates the production of hydrogen-sulfide during fermentation. An earlier study did not find any metal effect on hydrogen-sulfide production. The reason for copper's ability to stimulate hydrogen-sulfide production is speculated to be either that the yeast produces more hydrogen-sulfide in order to neutralize the excess copper by precipitating it as copper sulfide, or that there is a natural selection of a copper-resistant yeast mutant whose characteristic is to produce higher concentrations of hydrogen-sulfide. The copper concentration that is of concern in this matter is in the 1mg/L range and above. In this connection it is interesting to note that in some European wineries there is an old tradition of putting a copper bar into the must vessel containing the white juice after crushing the grapes. This is done to introduce some copper ions into the must, so that any hydrogen-sulfide that might be formed during fermentation will react immediately with the copper ions, and precipitate as copper sulfide. This preventative action (to eliminate potential stinking of the wine) promotes more production of the agent that it is aimed to remove. It should be emphasized that only copper addition *before* fermentation may facilitate hydrogen-sulfide production. Addition of copper after fermentation is finished, in order to

reduce an excess of hydrogen-sulfide, has no such effect.

Mercaptan (thiol): When hydrogen-sulfide is formed at high concentrations and is not removed from the wine, it may react with ethanol to form *ethyl-mercaptan* CH_3- CH_2- SH and *diethyl-sulfide* CH_3- CH_2- S - CH_2- CH_3. (Mercaptans are also called ethanethiols).

The odor thresholds of ethyl-mercaptan and diethyl-sulfide are 1 μg/L. This value is very low in comparison to other off-odor substances, and it is very small compared to their precursor, hydrogen-sulfide, which is (50 – 100) μg/L.

If hydrogen-sulfide was produced during fermentation (in the case of nitrogen deficiency or due to other reasons) it should be removed from the wine as soon as possible. The simplest way to do it is by aeration (and some oxidation), namely, the whole volume of the wine is pumped from one tank to another (or back into the same tank) while the wine is splashed over the top of the receiving tank. In most cases this measure suffices, and the "stinking" wine loses its off-odor right after this operation. In more severe cases, mainly when mercaptan has been formed, chemical treatment is needed to remove it. To treat such wine, copper addition might help, up to 0.3 ppm of copper. The legal limit for copper addition is 0.5 ppm. Over-fining with copper may also cause copper haze. If addition of copper reduces the bad smell, but not completely, carbon treatment (see the Cellar Operation chapter, section C) might be of great help. In any case, a laboratory test to find the minimal amount needed is necessary before any addition of copper is made.

Phenolic Haze: This is very rare, although it may happen in white wines. Two types are known:

(1) *Flavonol haze:* Seen in recent years, in wines treated by the regular protein stability treatment, which then developed haze. It was shown to consist of the flavonols-protein complex. The specific flavonols in such haze formation (*quercetin, kaempferol, myricetin*) are found in quite low concentrations in grape juice. After being hydrolyzed (from their glycosides) they may interact with some proteins left over and form yellow sediment. The source for these flavonols was attributed to grape leaves found in the fermenting wine due to mechanical harvest, which introduces higher leaf content into the crush than hand harvest.

(2) *Ellagic-acid:* This was observed in white wines treated with oak chips. A brown deposit was found in filtered, bottled wines, a few weeks after bottling. The bottling was done in a short period after the wine was treated with fresh oak chips. The deposit was identified as *ellagic-acid*, which was precipitated from the solution

as a result of the hydrolysis of ellagitannin glycosides. Ellagic-acid is insoluble in a wine medium. The ellagic-acid precipitation was confirmed in model solutions containing 10% alcohol and saturated with potassium bitartrate. The solution was treated with oak shavings, then filtered and stored. A deposit occurred after a few days and again was verified as ellagic-acid. The phenomenon of ellagic-acid precipitation probably exists in red wine too, but was not noticed because of the dark color and the normal expectation of a tannin deposit in red wines. In white wines aged in barrels (or treated with oak chips), where the wine is bottled soon after being removed from the wood, such a brown deposit is a serious potential risk. It takes a few days to a few weeks for the deposit to be formed, so it may be advisable to delay the bottling for few weeks and then bottle the wine after membrane filtering. Then, the wine might be free of this problem.

Over-Processing: This term includes operations during vinification that have been done carelessly, or without attention to the impact on wine quality. In this term we can include such operations as filtering, fining, racking, sulfiting, and oak aging.

(1) *Over-filtering:* Too much pad filtering, or with too much DE powder, may give the wine a "papery" or "earthy" taste. Red wines should not be filtered at all during processing, except before bottling with a coarse or fine filter, but can also be bottled without any filtering, as some wineries do. The best clarification of red wine that is aged in barrels is by natural precipitation. White wines can be filtered once or twice after cold stabilization and, finally, sterile bottle filtration.

(2) *Over-Fining:* Fining with any agent (such as bentonite, gelatin, carbon, etc.) should be done with the minimum quantity required to fulfill its function; otherwise it will "strip" the wine of its aroma and flavor.

(3) *Over-racking:* This will introduce too much oxygen into the wine, causing over-oxidation. Too much oxygen in white wines may show up as a yellowish color, lack of fruitiness, flat aroma, bitterness, and even a caramel flavor. In rosé wines or light reds, the color will be a pink-brown or red-brown and is usually accompanied by a flat aroma. Red wine pigments are oxidized and polymerized to lose their red color. They would age in the bottle quite rapidly, depending on the amount of tannins in the wine and their over-exposure to oxygen. In order to prevent this, minimum aeration should be practiced during wine processing, especially in white and "blush" wines. In general, red wines are less susceptible to oxygen, hence less vulnerable to racking operations.

(4) *Over-sulfiting:* Sulfur-dioxide is added to the wine during its processing from

crushing to bottling. These additions are cumulative, and at the end, the wine in the bottle will contain a certain quantity of bound SO_2 (which is difficult to recognize by taste) and a certain amount of free SO_2, which is described as a burning sulfur or pungent smell. Excessive free SO_2 rarely happens in modern winemaking, as the SO_2 regime is strictly controlled. However, if it does happen (probably by mistake), the way to try to reduce the excess free SO_2 from the wine is to strip it by *aeration* or by *oxidation*. The aeration stripping can be done either by racking (from the bottom of the tank to the top of another one) or by bubbling nitrogen through the wine for a couple of minutes (from the bottom of the tank). The aeration method is not efficient enough, but it is the minimum effort to try to reduce the excess.

The other option (oxidation) is more efficient and is based on the oxidation of SO_2 with hydrogen-peroxide: $SO_2 + H_2O_2 ===> SO_4^= + 2H^+$
To do it, follow this guideline: In order to reduce by 10 ppm the free SO_2 in the wine, add 50 ml of 1% hydrogen-peroxide to 1 HL of wine. An excess addition of peroxide may oxidize other components in the wine, reducing its quality. Therefore, the addition should be made only after a small-scale laboratory test has been performed to find the exact volume of peroxide needed to reduce the SO_2 to the desired level, without affecting the wine quality. The addition should be done in small steps, 10 ppm at a time.

(5) *Over-barrel aging:* This may cause the wine to taste and smell overly oaky, which sometimes dominates the natural wine aroma and flavor. The wine tastes unbalanced and its quality is definitely reduced. Tasting the wine during barrel aging is necessary to prevent this.

2. Microbial Faults

Microbial faults are defined as the changes that occur in wine, as a result of microorganism growth which causes chemical changes in wine composition. The phenomena of wine getting spoiled are well known in wine production, probably for as long as wine has been made. These changes may appear as turbidity, sediments, gassiness, color change, and sensory changes (off-odor and off-taste). Wine is not a favorable medium for microorganisms to grow in. It contains alcohol at about (10 – 14) %, its pH is quite low (3.0 – 4.0), and it contains various compounds that are not favorable to microorganisms to live in, such as sulfur-dioxide, higher alcohols, phenols, and others. In spite of these difficulties, some "bad" yeast and "bad" bacteria are able to find their specific substrate in wine and are capable of growing in it, if the right conditions are met. When such a harmful microorganism is able to grow in wine, we say that the wine is "sick." When the results of this sickness are noticed, usually it is too late to cure the damage that has occurred in the wine (in the bottle, and in some cases even before). For convenience let us divide the various microorganisms that spoil wine into two categories, namely, *aerobic* (which need air for their activity), and *anaerobic*.

Aerobics Spoilage

Acetic acid bacteria – These bacteria are very well known for causing wines to change to vinegar. We should not forget that this problem was so common in the past that on the scorecard for wine evaluation developed in Davis, California in the 1950s (60 years ago), the wine was given a 10% score *just for not having volatile-acid* (see section F in this chapter). Such credit was abandoned some years later because "sick" wines containing high volatile acidity became quite rare. Acetic acid bacteria are classified into two genera, *Acetobacter* and *Gluconobacter*. Both genera make their living by oxidizing ethanol to acetic acid, but the difference between them is that Acetobacter are able to proceed further with the oxidation to CO_2 and water, while Gluconobacter cannot. Acetobacter include four species: *Acetobacter-aceti, Acetobacter-liquefaciens, Acetobacter-pasteurianus* and *Acetobacter-hansenii*. The Gluconobacter genus includes one species, *Gluconobacter-oxydans*.
Several factors affect the survival and growth of acetic acid bacteria in wine:

(1) *pH* – The optimum pH range for their growth is at about (5.0 – 6.5), which is much higher than the wine pH. The lower tolerance limit is at pH = 3.0, which is actually the lower pH range of wines.

(2) *Ethanol* – Although ethanol is their substrate, acetic acid bacteria are sensitive to its concentration. A wine ethanol concentration of (10 – 14) % is quite high for their survival, although they were found to grow in wines up to 15% ethanol, their uppermost limit.

(3) *Temperature* – Optimal growth temperature is about (25 – 30) °C. No data are available on the lower and upper temperatures that inhibit their growth. At room temperature range, their growth rate is increased as the temperature is raised.

(4) *Oxygen* – Oxygen is vital for their growth. Any exposure of the wine to air may increase their activity, especially on the surface of stored wine in tanks or in barrels. As a matter of fact, the greatest risk of these bacteria is during barrel aging with a regular supply of oxygen. There is a correlation between oxygen concentra-

tion in the barrel and acetic-acid production by *acetobacter*, which is more tolerant to high alcohol concentration. (The *Gluconobacter-oxydans* species cannot grow in barrels in a high ethanol concentration over 10%.)

Above 50% oxygen saturation, *acetobacter* is very active to oxygen, and below 30% its activity is almost completely halted. Thus, after any short exposure of the wine, such as racking, filtering, etc., there is a risk that the acetic acid concentration will increase. In a survey on the occurrence of acetic acid in Australian wines covering over 7000 wine samples, it was found that in white wines the mean value was 0.5g/L, and in red wines 0.65g/L (with 32% of them over 0.7g/L). At such a VA value (0.7g/L) the wine is still under detection threshold, but not far from it. The VA might be from any source during vinification, but it shows how this problematic factor is still common in the industry, although very much in control.

(5) *Sulfur-dioxide* – This has almost no effect on the acetic acid bacteria growth in the regular concentrations of SO_2 in wine. Free sulfur-dioxide in the 25 ppm range is not enough to control their growth. More information is needed in this matter.

The overall reaction that takes place when these bacteria are present is:

$$CH_3\text{-}CH_2\text{-}OH \xrightarrow{[O_2]} CH_3\text{-}CHO \xrightarrow{[O_2]} CH_3\text{-}COOH$$

The oxidation from ethanol to acetaldehyde and then to acetic acid is achieved by the use of atmospheric oxygen, which makes it a prime condition for the bacteria to proliferate. The acetic acid formed by these bacteria is added to the existing acetic acid and other volatile acids formed by the yeast activity. The legal limits of volatile acids in the US are 1.2 and 1.4 g/L in white and red wines respectively. In a normal fermentation the VA concentration range is (0.3 – 0.5) g/L range.

The odor threshold levels of acetic acid in white and red wines are 1.2 and 1.0 g/L respectively. Wine containing acetic-acid at a concentration above (2 – 3) g/L is unpleasant, and is considered spoiled. A more disturbing compound associated with this spoilage is the ethyl-acetate (CH3-CH2-O-CO-CH3) that is always formed by the acetic-acid bacteria as a byproduct accompanying the formation of acetic acid. The odor of spoiled wine infected by the acetic acid bacteria is mainly caused by this ester rather than by the acid itself. Its odor threshold level (in white and red wines) is in the (150 – 200) mg/L range. When wine is infected by acetic acid bacteria beyond the spoilage level, there is not any known treatment to cure it. The best measure for avoiding acetic acid bacterial growth during the winemaking process is never to leave any wine stored in a container that is not full, or if that is impossible, to cover the top of the wine with a nitrogen or carbon dioxide blanket.

Flor sherry – This contains various genera of aerobic surface film yeasts that grow on the wine surface when air is available, and consume its alcohol to produce acetaldehyde as their oxidation product. They form a white film on the wine surface which is the result of incomplete separation of the dividing cells after budding. The cells have a cylindrical shape with dimensions of about $(2 - 4)$ μ wide, by $(5 - 10)$ μ long. These yeasts belong to the genera *Candida* (mostly *Candida-mycoderma* species), *Pichia* (mostly *Pichia-membrafaciens* species), *Hansenula* (mostly *Hansenula-anomala* species) and *Saccharomyces* (mostly *Saccharomyces-beticus* and *Saccharomyces-fermentati* species). In vinification of table wines (not sherry type wines), if these yeast infections are left on the wine surface for a certain length of time, they may spoil the wine with the undesirable acetaldehyde odor. On the other hand, they are encouraged to grow on sherry wines, where the acetaldehyde gives these wines their special and typical character. Certain conditions have to be met for this yeast to grow. They are inhibited by a high level of alcohol, above $(12 - 13)$ %, and by low pH and sulfur-dioxide. In most cases the infection occurs in the un-topped tank, but sometimes the yeast may grow to some extent in the bottle, as a result of non-sterile filtration, air exposure during bottling, and not enough free SO_2.

Anaerobic Spoilage

Fermentation yeast – In wine that is not completely dry and containing more than $(1.5 - 2.0)$ % fermentable sugar, and when suitable measures have not been taken (sterile filtration, sorbate, SO_2), re-fermentation will probably occur. It may happen during wine processing or in the bottle, where the damage is the most harmful. When it happens in the bottle, it may cause some gassiness and sediment in the easy cases or even bottle explosion in the severe cases.

Yeast can also produce acetic acid as a by-product of fermentation. The cultured yeast of *Saccharomyces-cerevisiae* and *Saccharomyces-bayanos* produces acetic acid in a normal range of $(0.3 - 0.6)$ g/L. But some "wild yeast" species (*Kloeckera-apiculata, Candida-stellata, Hansenula-anomala*) that exist on the grape skins are very productive of acetic-acid, mainly at the early stage of fermentation before the inoculated yeasts take over. The risk is more severe with "natural fermentation." Also, at lower temperatures of fermentation (at about 15°C) these indigenous species have a higher growth rate than the *Saccharomyces* yeasts, increasing the risk of higher VA concentration. This is more common in white wine fermentation at low temperatures.

Malolactic-acid bacteria – This is a very important microorganism factor in winemaking. (More detailed information can be found in Chapter III, section D.)

In this section it will be presented only from the potential spoilage point of view. The wine is spoiled by these bacteria if malolactic fermentation occurs in the bottle. The major reaction that takes place when these bacteria are present is the fermentation of malic acid to lactic acid, releasing carbon dioxide. Other by-products (2,3-butandiol, diacetyl, acetoin, acetic acid, and others) are also produced; some are desirable (buttery-like odor), and others are detrimental to the wine.

Malolactic-bacteria are divided into three genera: *Leuconostoc*, *Pediococcus*, and *Lactobacillus*. The commercial cultivated species that is used for induced malolactic fermentation is *Leuconostoc-oenos* (the old name) which is now called *Oenococcus-oeni*. It does its job rapidly and without undesirable off-odors. The commercial species are also able to produce acetic-acid, but in very moderate concentrations of up to 0.5g/L, compared to the other genera which are able to produce acetic-acid up to a few grams/L. The other genera may also cause some other problems. *Pediococcus* and *Lactobacillus* are slower fermenters, which leaves an unpleasant vegetative smell, mainly if the fermentation takes place in the bottle. Lactic acid bacteria can use other compounds as their substrate such as carbohydrates (glucose and fructose), even in wine containing less than 1% sugar. In fermenting wine, the risk is significant in cases where the wine was inoculated with malolactic bacteria and the fermentation became stuck. The ML bacteria take over the lead in using the sugar to produce mainly acetic acid. The VA of such stuck fermenting wine becomes much higher than in normal fermentation. Immediate re-fermentation is vital in order to save the wine.

When sorbic-acid has been added to the wine in order to preserve it from yeast growth (when some residual sugar is left over), malolactic bacteria may utilize this substrate to form *2-ethoxyhexa-3,5-diene*, which has a very distinctive geranium-like smell.

Lactic acid bacteria can also ferment tartaric acid, transforming it into lactic and acetic acids. The essential conditions for such sickness are low acidity and low sulfur-dioxide. The results are reduction in fixed acidity, increase in volatile acidity, cloudiness, gassiness (carbon dioxide), and off-odors. In the old French literature this "sickness" of wine was called *vin tourné*. To avoid the risk of malolactic fermentation in the bottle, it should either be encouraged to take place under controlled conditions during the wine processing in the cellar (mainly in red and white wines that are intended to age), or be prevented by taking certain measures that are detailed in Chapter III (low pH, SO_2, inhibitors such as fumaric-acid and sterile filtration).

Brettanomyces/Dekkera – The double name of these yeast genera refers to different forms of the same microorganism. The asexual form (budding reproduction) is called *Brettanomyces*, while the sexual form (sporulation) is called *Dekkera*. This yeast is found in wines and in beers. Not enough is known about this yeast, namely its metabolic and growing conditions. It can grow in certain wines, and not in others, but we have no clear understanding as to why. Red wines are its favorite. A low SO_2 concentration does not inhibit its growth, nor does sorbate. About 25 ppm of free SO_2 is needed for temporary inhibition. When the free SO_2 becomes bound and reduced to a low level, it may resume its activity. Probably the most effective method to prevent its growth is sterile filtration. This is not always practical, at least in cases where red wines are not filtered (or are filtered just by a coarse filtration) in order to minimize wine stripping. The distribution and infection of this yeast appears to be local in certain wineries, and it is transmitted within the winery through the equipment, barrels and tanks.

The major contribution of this yeast to wine flavor is the production of *volatile phenols*: *4-vinylphenol, 4-ethylphenol, 4-vinylguaiacol* and *4-ethylguaiacol*.

They are derived from the two cinnamic-acid-phenolic, *p-coumaric-acid* and *ferulic-acid*, by enzymatic degradation:

| p-coumaric-acid | 4-ethylphenol | ferulic-acid | 4-vinylguaiacol |

Only these two phenolic-acids (*p-coumaric* and *ferulic*) are enzymatically decarboxylated. The concentrations of these compounds in red wines were found to be in ranges of (tens to hundreds) μg/L, in many cases much above their thresholds. These compounds are responsible for the *animalic, leather-like, barnyard, mousy, wet blanket*, and other such descriptions.

(Malolactic bacteria are also able to produce these volatile phenols, but with much lower production compared to Brettanomyces yeast. It was found that some "wild" yeast species are also able to de-carboxylate *ferulic-acid* to *4-vinyl-guaiacol*, contributing its off-flavor.)

The organoleptic effect of Brettanomyces infection on wines is also enhanced by production of two other compounds, described also in terms like those above. The "mousy" odor was related to the formation of two heterocyclic compounds: *2-acetyl-1,4,5,6-tetrahydropyridine* and its *tautomer 2-acetyl-2,4,5,6-tetrahydropyridine:*

2-acetyl-1,4,5,6,-tetrahydropyridine   2-acetyl-3,4,5,6,-tetrahydropyridine

These compounds are known to be components in baked bread, with an odor threshold of 1.6 ppb. It was shown that the biosynthesis of these compounds is involved with the amino acid lysine and ethanol as precursors.

All the above compounds produced by *Brettanomyces* in low concentrations add to the wines some kind of "complex" nose, which then, as the infection advances, turns out to be a fault.

And now, why are the Brett characteristics are found only in red wines? Very simply, because the precursors (phenolic-acids) for producing enough volatile-phenols, are found in red wines.

The conditions which are most suitable for Brett growth are met in the barrels (their favorite place to grow). These conditions are: wine with high pH (above 3.6), sufficient oxygen supply, higher temperatures (above 20°C) and low concentration of SO_2 below (15 – 20) ppm free. And even under good conditions, it grows slowly, without revealing its potential risk. Brettanomyces yeast can also produce a significant amount of acetic acid at levels of a few g/L, adding its share to the accumulation of this fault agent in wine.

In order to control the Brettanomyces growth in wine, two measures can be taken. First, maintain good sanitary conditions in the winery and in any wine container, especially barrels, which are the site of its main activity during vinification. Second, it is very important to maintain a high free SO_2 level during all stages of production. A weak point during wine production is the interval between ending the alcoholic fermentation and finishing MLF. At this time the SO_2 level must remain low, in order not to inhibit the ML bacteria. During this period, which may take a few weeks or more, the wine is vulnerable, and the Brett yeast can take hold in the wine with long-term results. It is therefore highly recommended that commercial ML bacteria be used to encourage MLF as soon as possible after alcoholic fermentation. Then, as soon as it is finished, SO_2 should be added. And one last note in this matter of controlling Brett growth—the oxygen. Exposure of the wine while in the barrels should be reduced if the winery is facing this problem. It was found that oxygen is a supporting factor in Brettanomyces activity, increasing its growth.

Saccharomyces-bailli – also called *zygosaccharomyces-bailii*, is unique yeast that can grow in very high sugar concentrations such as grape juice concentrates with (50 – 70) % sugar, which means that they can tolerate very high osmotic pressure. They are also highly resistant to sulfur-dioxide, to sorbic-acid and even to *dimethylpoly-carbonate* (DMPC). The yeast shows good resistance to all three agents at a concentration range of up to 200 ppm. This yeast is quite rare and was found in sweet wines where their source of sweetness came from grape juice concentrate. Being so unaffected by various inhibition agents, the only measure that can preserve an off-dry wine from being spoiled in the bottle by this yeast is sterile filtration.

3. Summary of Faults
General Comments

First, let us distinguish between a *wine flaw* and a *wine fault*.

A wine *flaw* is a small diversion from the common acceptance of how wine should be made. It does not deal with wine quality, but with some technical diversion in wine composition that can be noticed by nose or taste or by sight as sediment or color change. Sometimes, in very low concentrations, around its threshold, it may even contribute some complexity to the wine and thus improve its quality. At higher concentrations it may decrease wine quality, ultimately causing a wine fault, namely, a very serious defect.

A wine *fault* is a major problem which is caused by sufficient concentration of a component that seriously reduces wine quality up to a point where the wine is undrinkable. All flaws and faults are caused by some inappropriate treatment of the wine during cellar operations or afterward in the bottle. The final reason is always a certain specific compound or compounds found in the wine. In some cases the flaw is caused by a completely innocent compound which happens to be in the wrong place at the wrong time, for example, some residual sugar above its acceptable threshold concentration in red wine. It is definitely not within the accepted norm of red wines, and in such a case it is considered a flaw. Too-high acidity in red wine may be unpleasant, especially when accompanied by high tannin, and both these components may cause a flaw in this wine. Very high tannin in young red wine may be considered a flaw which will completely vanish in a few years, so it is sometimes a temporary flaw. Some phenolic may be perceived as bitter and unpleasant; that constitutes a flaw. As a matter of fact, anything that causes wine to be unpleasant in its smell or taste, temporarily or permanently, is a flaw. On the other hand, acetic acid, for example, at a concentration slightly above its threshold, is so serious and unpleasant that it is considered a fault. But at the same time, this acid in late-harvest wine could be at much higher concentration than its threshold and still be considered okay. This is so because of the very high sugar content which masks the acetic sensation, and it is accepted as a norm in late-harvest wines.

There are many such examples, with the main message that the definition of a flaw or a fault in wine is very flexible, and sometimes temporary. But there are some compounds in wine that have a very high organoleptic impact, for good or for bad, at low to very low concentrations. And in some cases the "good" and the "bad" sensation depends on their concentration. For example, the group of compounds related to Brettanomyces may have a pleasant impact at very low concentration, which turns out to be unpleasant at a higher concentration. Another example is *diacetyl*, which is one of the by-products of MLF (and of some alcoholic fermentation as well), which adds some "buttery" flavor to the wine at a low concentration in the few mg/L range, and becomes an off-flavor of rancid butter at over 5 mg/L.

Tables of Flaw/Fault Compounds

In the following tables we summarize the most-known specific compounds which cause wine faults or flaws. The tables include the compound name and structure, its detection threshold, its organoleptic description, and its possible source.

Some of the faults are caused by more than one source. Acetic-acid is probably number one in being produced by many agents, and its final concentration is the accumulated sum of all of them. It is produced all along the vinification process, starting at alcoholic fermentation by yeast (cultured and "wild"), by *Malolactic* bacteria *Oenococcus-oeni* (the "good" ones) and by *Pediococcus* and *Lactobacillus* (the "bad" ones). And then it continues in the barrels by *Brettanomyces* yeast, and lastly by *Acetobacter* bacteria.

Volatile-phenols are also produced by various agents, namely, *Brettanomyces* and *Malolactic bacteria*. They are produced by chemical and microbial reactions in the barrel's wood and in the cork. Also, their smell is defined in so many terms that depend on their actual concentration in that specific case.

The threshold concentrations are given in the literature in various units which are sometimes confusing, such as mg/L, μg/L, ng/L, ppm, ppb, ppt. In order to bring some common basis to all these threshold units, here are the definitions of the common units (each unit jumps in increments of 10^{-3}):

1 ppm = 10^{-3} gram/L = 1 mg/L (milligram/liter)

1 ppb = 10^{-6} gram/L = 1 μg/L (microgram/liter)

1 ppt = 10^{-9} gram/L = 1 ng/L (nanogram/liter)

In the table we will use the three units: **mg/L, μg/L, ng/L**.

The threshold values in the literature for a specific compound are given in many studies in quite wide ranges, and careful selection was made to judge which are the most reliable. In any case they give a good orientation at what concentration range the compound occurs. The compounds are divided in groups according to their common source in wine.

Sulfur related components:

Compound	threshold	description	source
Hydrogen-sulfide H_2S	10 / 80 µg/L Air / wine	rotten eggs	fermentation reductive processes
Methanethiol Methyl-mercaptan CH_3-SH	0.3 µg/L	cooked cabbage	low molecular weight thiols
Ethanethiol Ethyl-mercaptan C_2H_5-SH	1 µg/L	onion, rubber	- " -
Dimethyl-sulfide CH_3-S-CH_3	30 / 50 µg/L white / red wine	cabbage, truffle	- " -
Diethyl-sulfide CH_3H_2 - S - CH_2CH_3	0.9 µg/L	garlic, asparagus	- " -

Oxidation related components:

Compound	threshold	description	source
Acetaldehyde CH_3-CHO	125 mg/L	sherry-like	oxidation
Acetic acid CH_3-COOH	0.8 / 1.2 g/L white / red wine	vinegar	acetobacter, ML bacteria, yeast
Ethyl-acetate CH_3-COO-C_2H_5	150 mg/L	acetone, varnish	acetobacter, yeast

Miscellaneous components:

Compound	threshold	description	source
2-Ethoxyhexa-3, 5-diene CH_3-CH(OEt)-CH=CH=CH_2	0.3mg/L	geranium	MLF + sorbate
Diacetyl CH_3-CO-CO-CH_3	1mg/L	rancid butter	MLF

Cork related components:

Compound	threshold	description	source
TCA 2-4-6-Trichloro-anizole 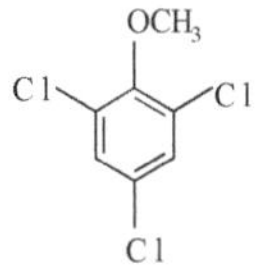 	4 - 10 ng/L	moldy, earthy cardboard	corks
1-Octen-3-one CH_3-[CH_2]$_4$-CO-CH=CH_2	20 ng/L	mushroom, metallic	corks
Geosmin Trans-1,10-dimethyl--trans-9-decalol 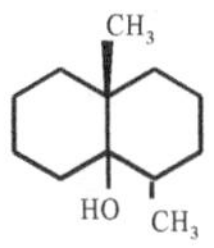 	25 ng/L	earthy	corks
2-Methyl-isoburaneol 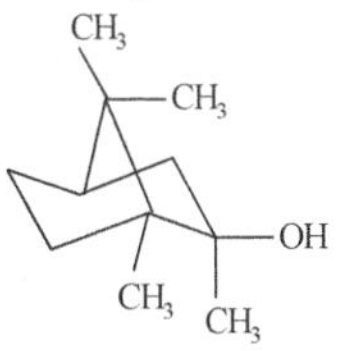 	30 ng/L	earthy	corks

Brettanomyces and MLF related components:

Compound	threshold	description	source
4-Ethylphenol 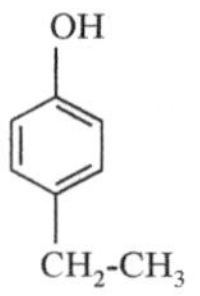 OH CH_2-CH_3	140 µg/L	barnyard, animal, wet horse, mousy	Brettanomyces MLF
4-Vinylphenol 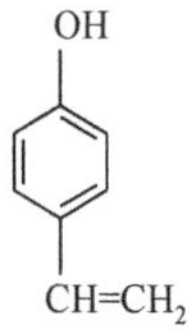 OH CH=CH_2	0.7mg/L	barnyard, animal, wet horse, mousy	Brettanomyces MLF
4-Vinylguaiacol 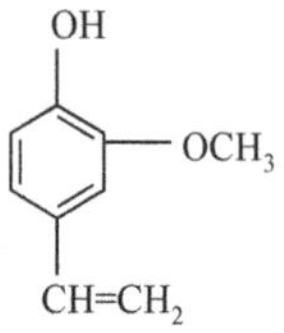 OH — OCH_3 CH=CH_2	200 µg/L	barnyard, animal, wet horse, mousy	Brettanomyces MLF
4-Ethylguaiacol 4-Ethyl-2-methoxy--phenol 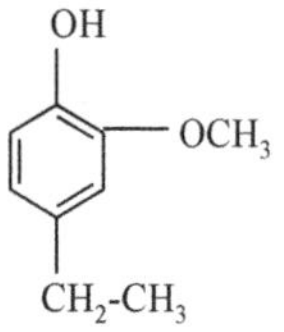 OH — OCH_3 CH_2-CH_3	100 µ/L	smoky, bacon, spice, animal	brettanomyces MLF
2-Acetyl-1,4,5,6--tetrahydro-pyridine 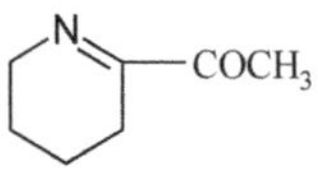 N — $COCH_3$	1.5µg/L	mousy	Brettanomyces MLF

Toasted oak and cork related components:

Compound	threshold	description	source
Guaiacol 2-Methoxyphenol 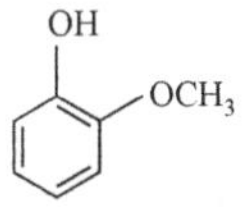 OH OCH₃	20 / 17 µg/L wine / water	smoky, burnt, medicinal	toasted oak corks
Eugenol 2-Methoxy-4- (2-propenyl) phenol 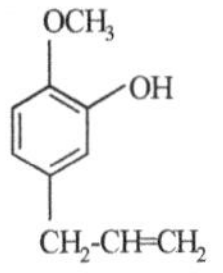 OCH₃ OH CH₂-CH=CH₂	0.7 / 0.2 mg/L red / white wine	spicy, clove	untoasted oak

E. Legal Regulations

Legal aspects of wine encompass a very wide range of activities such as establishing a winery, production, trade, taxes, and others. In every country that produces wine there are wine laws and regulations which are tailored to the specific characteristics and needs of the people living in that country. Most regulations are supposed to protect the customer against risks to his health and pocket, and in return, to collect money (by taxes), back from that pocket. One should not forget, also, that many administrative jobs are produced to maintain and supervise wine regulations. Like any other laws, wine laws reflect the attitude and feelings of that society toward wine and wine culture. By focusing on one country, a good example of the concept and an introduction to this subject can be gained. Technical regulations such as production procedures, wine content and wine additives are more or less quite common to most countries. The focus here is, naturally, the chemical aspects of the law, but several other topics in this area will also be discussed.

Basically, wine laws cover three phases of winemaking, namely, viticulture practice, wine production, and wine trading. Viticulture laws regulate the appellation designation, the varieties, crop yield, and grape trade. Wine production laws control the issues that concern public health and fraud prevention. Trade laws monitor the marketing system and the tax collection. In this section we shall view the legal aspects of wine through the regulatory system in the United States. In some respects, this country has some very unique regulations, but the technical regulations are basically similar to any other country. In the U.S., federal wine laws are under the jurisdiction of the Department of the Treasury, and are enforced by a sub-unit, the Alchohol and Tobacco Tax and Trade Bureau, abbreviated TTB. This office is in charge of the production of wine and collecting money (as established by the Internal Revenue Code). The distribution, labeling and advertising of wine is controlled by the Federal Alcohol Administration Act (FAA). Besides federal regulations, each state has its own control (and taxes) on wine production, appellation and sale. The current federal regulations are a revision that was enacted in December 1990. They are updated to the technological changes that have happened in the wine industry during the last decades.

1. Definitions

Grape wine – Wine produced by the normal alcoholic fermentation of the juice of sound, ripe grapes. Minimum alcohol content is not less than 7% (v/v).

Fruit wine – Wine produced by the normal alcoholic fermentation of the juice of sound, ripe fruit. Minimum alcohol content is not less than 7% (v/v). If the wine is made from one kind of fruit, it should be called wine with the fruit name (e.g., *berry wine*). If the wine is made from apples, it may be called *cider*. If the wine is made from more than one kind of fruit, it should be called *fruit wine*.

Agricultural Wine – Wine produced by the normal alcoholic fermentation of agricultural products other than grapes and fruits (e.g., honey, but not grain or molasses).

Table Wine – Grape wine with an alcohol content not in excess of 14% (v/v).

Dessert Wine – Grape wine with an alcohol content between (14 – 24) %. The alcohol added to fortify the dessert wine should be grape alcohol (brandy). Specific wines in this category are *sherry*, which should contain not less than 17% alcohol. *Port* wine should contain not less than 18% alcohol. In a case where the alcohol content of sherry wine is between (14 – 17) % and port wine between (14 – 18) %, these wines should be called *light sherry* or *light port*.

Sparkling Wine – An effervescent wine containing more than 3.92 g/L of carbon dioxide, resulting solely from secondary fermentation of the wine within a closed container.

Champagne Method – This is defined as sparkling wine whereby it derives its effervescence solely from secondary fermentation within containers of not greater than one gallon, and which possesses the taste, aroma, and characteristics attributed to Champagne wines. In most countries other than the U.S. such protected names (as Champagne) are not allowed. Sparkling wines made in larger containers (mostly pressure tanks) cannot use the term "Champagne method," but rather the term *Charmat method*. If the wine is charged by artificial carbon dioxide, it is called *carbonated wine*.

Appellation – Two definitions: *American wines*, and *American viticultural areas*. The first term specifies the state and county where the grapes are grown. The second term denotes specific viticultural areas which have been approved as such. This appellation term does not necessary overlap the geopolitical boundaries definition, but it is defined as a specific grape growing region approved by the TTB. Sonoma County in California for example, consists of eleven viticultural areas. Unlike other countries, the U.S. appellation system does not indicate grape variety or viticultural methods control, and no quality criteria or characteristics of the wines made in the viticultural areas are required. The federal regulations require that at least 75% of the grapes must be from the state and county which are printed on the label. In California, 100% of the grapes must come from California in order to use the state's name (California wine) on the label. For a *viticultural area* within California, at least 85% of the grapes should be from that area.

Geographical semi-generic names – Names that indicate specific characteristics of the wine and not their geographical origin. For example, Champagne, Port, Sherry, Chablis are considered and recognized by their type of wine and not as having been imported from abroad.

Estate bottled – May be used on the label if the wine is labeled with a viticultural area appellation and several other conditions are met: The bottling winery is located on the land where the grapes are grown, all the wine is made from grapes grown on this land, and the whole process of making the wine, including crushing, fermenting, aging, and bottling is done at that winery. No other term which may indicate any connection between winemaking and growing the grapes is allowed.

Varietal wine – The wine should contain at least 75% of the grape variety whose name is printed on the label. A blend of wine that contains less than 75% of one variety should have a generic name. Of course, any varietal wine (by the above definition) can also be called by a generic name.

Generic wine – Wine not made from one dominant grape variety that is at least 75% of its content, and hence cannot be called a varietal wine.

Vintage wine – Wine that contains at least 95% of grapes harvested in the year identified on the label.

2. General Regulations

Wine label – Shall not be false or misleading. It should contain all the following information:

(1) The name of the producer or the brand name.

(2) The bottler name and address.

(3) The kind of wine (e.g., varietal or generic table wine, dessert wine, etc.).

(4) The vintage.

(5) The alcohol content (v/v), except in table wines, where the alternative option is to write the words "table wine," which automatically indicates that its alcohol content is in the (7 – 14) % range.

(6) The bottle content in metric sizes. The most common bottle size is 750 ml in almost every country in the world. Other sizes are multiplications of the basic

750 ml, namely, 375 ml, 1.5 L, 3.0 L, 6.0 L and 12 L. Due to lack of uniformity in bottling, the volume of wine in the bottle shall not be *lower* by 1.5% in 1.5 L bottle (22.5 ml), 2% in 750 ml bottle (15 ml), and 3% in 375 ml bottle (11 ml).

(7) Shall contain certain warnings in regard to health risks such as alcohol use during pregnancy, sulfite content (if wine contains more than 10 ppm), and also the warning that alcohol may impair the ability to drive a car or operate machinery. The wine label must be approved by the TTB.

Taxes – are paid according to federal *and* state regulations, by the wine producer at the time of removal of the wine from the winery. The taxes are progressively increased (based on alcohol content) according to the following categories:

(1) Table wine containing not more than 14% alcohol.

(2) Wines containing more than 14% alcohol and less than 21%.

(3) Wines containing more than 21% alcohol and less than 24%.

(4) Artificially carbonated sparkling wines.

(5) Sparkling wines.

When the alcoholic product contains in excess of 24% alcohol, it is taxed in the category of *distilled spirits*.

3. Wine Content

The following materials in this section are those which are controlled during wine production, either as additives to the wine or as having been included in the wine during its processing. The wine *treatment additives* are not included here, but only the basic materials that are considered as *wine components*. The wine contents are presented in the following section.

Water – May be added to the must before fermentation, providing that the sugar content of the must will not be reduced below 22 Brix. This practice is not permitted in California. Instead, the unclear and ambiguous regulation in California states that "no water in excess of the minimum amount necessary to facilitate normal fermentation may be used in the production or cellar treatment of any grape wine." What do you think this means? Is it allowed or not? In practice, wineries add water to the must when the Brix is high (say 25 and above), in order not to have wines with higher alcohol content (to avoid taxation and "hot" wine).

Sugar – In case of low sugar content, dry sugar or concentrated juice may be added before or during fermentation. The sugar addition (*chaptalization*) should not raise the juice density above 25 Brix. Addition of (17 – 19) gram of cane sugar (sucrose) per 1L of must will increase the alcohol content by 1%. This formula can serve as a general guide for the quantity of sugar needed. There is no definition as to what is a low sugar content, so in practice, sugar addition is permitted in the U.S. In California, no sugar other than pure grape concentrate can be used (before or after fermentation) in the production of still wine. It is permissible in production of sparkling wines before the secondary fermentation (dosage).

Acids – Tartaric and/or malic acids can be added to must before fermentation to correct acid deficiency. After fermentation, other acids such as citric, fumaric, lactic (besides tartaric and malic) can be added up to a fixed TA limit (calculated as tartaric) of 9 g/L. In addition to the maximum TA limit, there are specific limitations on the content of each acid in wine as follows: citric-acid 0.7 g/L and fumaric-acid 2.4 g/L.

Volatile acids – This is mainly acetic acid. The current limits are 1.2 g/L and 1.4 g/L in red and white wines respectively.

Hydrogen-sulfide – There is no limitation on the content of this component in wine.

Methanol – The wine shall not contain more than 1.0 g/L.

4. Wine Treatment Additives

A summary of the most useful materials used in wine production is given in the following tables (the word GRAS in the tables mean Generally Recognized As Safe by the FDA). For full information on the usage of these substances, see chapter VI on Cellar Operation, section C.

Yeast nutrient materials:

Material	Used for	Limitations
Di-ammonium-phosphate	Phosphate and nitrogen yeast nutrients	GRAS. Not to exceed 960 mg/L.
Ammonium carbonate	Nitrogen yeast nutrient	GRAS. Not to exceed 240 mg/L.
Thiamine	Vitamin yeast nutrient	GRAS. Not to exceed 0.6 mg/L.
Yeast extract	Yeast nutrient	GRAS. Not to exceed 360 mg/L.

Preservative materials:

Material	Used for	Limitations
Ascorbic acid & its isomer Erythorbic acid	Anti-oxidant agent	GRAS
Dimethyl dicarbonate	Antimicrobial agent	Not exceed 200 mg/L
Fumaric acid	Inhibit ML fermentation	Not exceed 2.4 g/L
Sorbic acid & Potassium Salt	To inhibit fermentation as sorbic acid.	Not exceed 300 mg/L
Sulfur-dioxide & other sulfite agents (sodium and potassium metabisulfite)	(1) Anti-oxidant agent (2) Antimicrobial agent	(200 – 350) mg/L in various countries

Acidity correction materials:

Material	Used for	Limitations
Calcium carbonate	To reduce excess acidity	GRAS. Acidity not to be reduced < 5.0 g/L.
Calcium sulfate (gypsum)	To lower pH	Sulfate not to exceed 2.0 g/L as K_2SO_4
Citric acid	To increase acidity	GRAS. Not to exceed 0.7 g/L.
Fumaric acid	To increase acidity	Not to exceed 2.4 g/L.
Lactic acid	To increase acidity	GRAS.
Malic acid	To increase acidity	GRAS.
Potassium carbonate & Potassium bicarbonate	To reduce excess acidity	GRAS. Acidity shall not be reduced below 5.0 g/L.
Tartaric acid	To increase acidity	GRAS.

Fining materials:

Material	Used for	Limitations
Bentonite (aluminosilicate)	Protein stabilization	GRAS. At 5% water slurry, not exceed 1% water of the wine vol.
Carbon (activated)	Fining agent for removal of color and off-odor	GRAS. Up to 3 g/L
Casein	To reduce tannin content	GRAS.
Copper sulfate	To remove hydrogen-sulfide	Not exceed 0.5 mg/L as copper, residual level not > 0.2mg/L
Egg-white (albumin)	To reduce tannin content	GRAS.
Gelatin	To reduce tannin content	GRAS.
Ferrocyanide (cufex)	To remove trace elements, sulfides, mercaptans.	GRAS. No residual level in finished wine in excess of 1 mg/L.
Isinglass	To reduce tannin content	GRAS.
Milk (pasteurized)	To reduce tannin content	Not to exceed 0.2% v/v of the wine
Polyvinylpolypyrrolidone (PVPP)	To reduce tannin content	Not exceed 7.2 g/L. should be removed by filtration
Potassium bitartrate	Cold stabilization agent	GRAS. Not to exceed 4.2 g/L.
Silica gel	Fining in combination with gelatin to reduce tannin	GRAS. Not to exceed 2.4 g/L as SiO_2 (30%) suspension agent

The following materials, which were permitted in the past for use as additives, have been delisted from the wine treatment materials: Ammonium carbonate, benzoic acid and its salts, hydrogen peroxide, propyleneglycol, polyvinylpyrrolidone (PVP), and urea.

F. Wine Evaluation

In order to evaluate wine and say something meaningful about it besides "I like it" or "I dislike it," a set of tools, concepts, and language has to be established. One of the major tools to evaluate sensory response to a stimulus (simple or complicated) is to test it by using people as the measuring instruments. In this section we shall touch very briefly the basic concepts of wine evaluation.

First, let us define some common terms used in wine evaluation:

1. Definitions

Aroma: This is a term used to describe the smell of a wine which can be contributed directly to the grape, namely, the primary nose of the grapes' components. Each variety and vineyard contributes a specific aroma (simple or complicated) to wine made from it, which characterizes its varietal uniqueness. Some varietal aromas are very powerful and easily recognized, while others are very weak and unspecific. Not only the grape variety is characterized by the wine aroma. The growing site of the grapes (terroir) may also contribute its uniqueness to the specific wine aroma.

Bouquet: This is the term for the secondary smell of wine, derived from its vinification and history. This term includes fermentation conditions, cellar operations and aging (barrel and bottle). The bouquet is continuously changing during the life of a wine, as it progresses from a young, fresh wine to an old, aged one. Bouquet is very sensitive to most cellar operations and to the storage conditions of the wine.

Taste: This is the tongue's response to sweet, salt, sour and bitter stimuli. Three of these are found in wine (sugar, acids, and bitter compounds) and are partially responsible for its taste perception.

Flavor: This is a general word used in wine evaluation language, in the narrow meaning as the "aroma sensation of the wine in the mouth," in parallel to its nose perception, called "aroma."

Body: This is a term that belongs to the "touch" perception, which is also known as "feeling." Touch is perceived by the whole mouth, and is stimulated by factors such as temperature, hot spices, burning due to alcohol, and tactile feeling due to dissolved gases, solid texture, and viscosity in liquids. In wine evaluation, two terms are of great importance, namely, *body* and *astringency*; both of them belong to this category of "feeling." The *body* is defined as the "fullness" or the "roundness" of the wine in the mouth. It is a sensation caused by the alcohol content, sugar content, and other dry extract compounds in the wine. This term spans the ranges from low body ("watery") to high body ("oily").

Astringency: This is the "squeezing" feeling in the mouth due to the presence of tannins. It is caused by the reaction between tannins and the proteins in the mouth, which creates a "rough" sensation in the mouth tissue.

Sensory evaluation tests: This is a wide field which studies the responses of humans to various stimuli and measures them by structured methods for this task. The field covers responses perceived by sight, sound, smell, and taste. For our subject we are interested in the last two. The methods were developed to deal with the very specific measuring instrument, a human being who is not calibrated in standard scale and is very individual in his responses. Therefore, training is the first thing to do in order to make any measurement, and second, statistical methods are needed to get something from the unclear raw data of an experiment. But this is the weakest point of these methods, because the results are an average which eliminates the edges.

Specifically for our interest, two major tests are very important and useful: *difference-test* and *threshold-test*.

The first one tries to give an answer to the question "Is there any difference between two samples?", and if yes, which is "more" or "less." The answer is not so simple if the difference is not obvious. The most effective and most frequently used tests in this category are the *Pair-test*, the *Duo-trio test*, and the *Triangle-test*.

The second major test, the *threshold-test*, tries to find out the various thresholds of sensitivity to specific stimuli by using the above methods. There are *detection thresholds* that find the lowest concentration of substance that can be *detected* as being different from a blank. And then there is the *recognition threshold* that finds the lowest concentration of a substance that can be *recognized* specifically, and not just be different from a blank. This threshold is always higher than the first one. Also, there is a difference in any threshold in different media, such as in water, in air, in white or red wine, etc. The results in the literature are very diverse and shed light on the real problem of dealing in these issues. (The reader who wants to find more details on this subject is referred to my *Concepts in Wine Chemistry* book, chapter IV.)

Going back to wine evaluation, there are two main aspects in evaluating wines. One is called *descriptive analysis*, and the other *rating tests*. The two aspects are presented in this section.

2. Descriptive Analysis

The object of this analysis is to create a common language and terms for describing the characteristics and quality of a beverage. The initiative to develop such common flavor vocabulary started in other food fields, while the major effort in alcoholic beverages was made by the brewing industry back in 1974-5. Terms had been defined to describe the possible flavors and sensations of beer. The terms were organized in a two-tier wheel, where the first was very general with terms such as "mouth feel," "aromatic," "nutty," "musty," etc. The terms in the second tier were more specific, for example "moldy," "leathery," and "papery" for the "musty" term in the first tier. The concept of an "aroma wheel" of beer flavor has become a useful tool in beer research.

At the same time a specific vocabulary of terms was developed for evaluation of cider flavor. These concepts were further followed by the whisky industry in Scotland in 1979. Four experts (two were skilled blenders and two sensory evaluation scientists) were assigned to "develop a common flavor terminology for use in the Scotch whisky industry," and "to improve channels of communication of flavor description." They developed a set of terms to describe the various characters of whisky at different aging stages. The result was a "whisky flavor wheel," which is presented on the following page. It contained two tiers. The internal tier is more general, describing a certain character, while the corresponding outer tier of the same section is more specific. For example, the inner term *phenolic* (hour 1 on the circle) is subdivided into *medicinal, peaty,* and *kippery,* or the inner tier term *sulfury* (hour 9 on the circle) is sub-termed as *stagnant, coal-gassy, rubbery,* and *cabbage.*

The wheel contains twelve major aroma terms and two flavor terms. The two flavors are *primary tastes* and *mouth feel effects.* They were subdivided into *sweet, salty, sour, bitter* (primary taste), and *mouth coating, astringent, mouth warming* (mouthfeel effects).

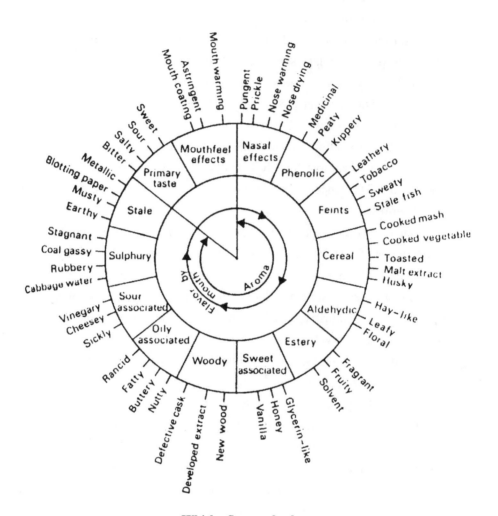

Whisky flavor wheel

The *woody* term which is so important in the whisky flavor received a special sub-wheel so that its very specific characteristics could be fully described:

By using scaling values to the different attributes describing certain food products, the beer industry adopted what was called "quantitative descriptive analysis," and multidimensional models representing products' characteristics were used. The strength of each attribute in the product is evaluated by a number in a given scale. The mean value of the panel represents its perceived effect among other attributes. The total attributes of a product's aroma perception has certain numbers which can give some descriptive knowledge about it. In order to represent these sets of numbers in an easier manner, the star figure was developed, where each line in the star represents an attribute, and its length represents its average value. All lines radiate from one point in all directions, forming a star-like shape. It has become very common to present the contribution of the multiple attributes of any product by this star figure. This presentation is widely used in the wine industry to present not only wines but also other related products, such as barrels, specific wine treatments, regional wine perception, brandies, whiskey, etc. The figure below is a polar coordinate graph of mean intensity rating ("a spider") for two competitive beer products:

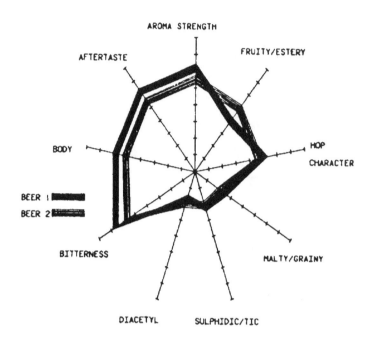

In developing such an instrument for describing whisky, the way was opened for further development of wine descriptive analysis with terms specifically related to wine attributes. In order to do descriptive analysis for wine, the various terms for describing wine must first be defined. These terms have to represent as fairly as possible the wine's domain of smells, flavors, and faults. Also, they have to be known and familiar to the taster's senses. Second, these terms have to be organized into groups, according to the associations of the members in that group or category. Then, the terms can serve as the vocabulary in wine evaluation language.

Such terms were defined, and standard solutions of components which can be used for training people were developed, in 1984-87 at the University of California in Davis. The results were published as what is called the "Wine Aroma Wheel," which contains in its inner cycle the very basic terms that may be associated with the wine under study, such as *fruity, vegetative, nutty, woody, earthy, floral* and others. The second cycle separates some of these terms into subgroups, for example, the *vegetative* is sub-grouped into *fresh, canned* and *dry*. The outer cycle is much more specific in describing the associated smell, e.g., *grass, bell pepper, mint, asparagus, green olive, hay, tobacco* and others. On smelling or tasting wine, it is possible in many cases to associate the *major* impression of the wine with some of the terms in the inner wheel, and maybe with some specific associations in the outer wheel as well. The use of these terms helps to communicate with other members of the tasting panel. The vocabulary can also be very useful for a wine critic to explain the wine's characteristics in words. The "Wine Aroma Terminology Wheel" is shown in the following figure:

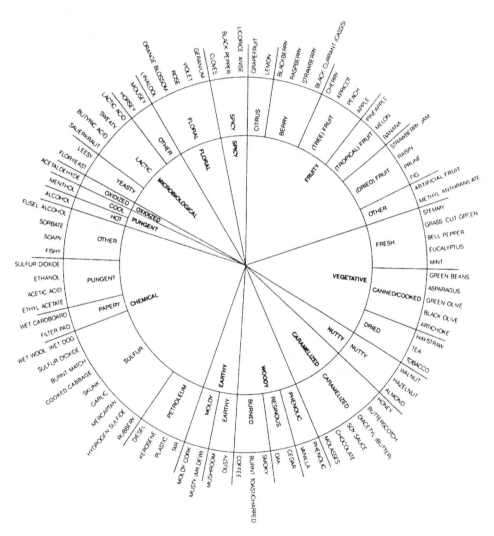

Wine aroma wheel

3. Rating Tests

In a *rating test* the question asked is, "On a given scale, where would you put each of the tested samples?" Before such a question can be asked, a set of basic rules has to be established. The first one is, what is the scale? The judge is asked to assign a number to his/her perception on the *quality* of a certain wine under study. The panel's collective numbers would produce a mean which represents the panel's judgment. In using such scaling techniques, some problems may arise. First, the individual perception of the judges on any issue can range widely. And then, on averaging, in most cases the result is a middle number which is rarely different from any other middle result. Because of the wide diversion of panel opinion, in most cases the *average* is the "big Viner," which means that very good wines get a lower score, and bad wines get a higher score than they deserve, with only a small difference between these wines. On a 100-point scale, almost all of the wines produced in the world are in the (85 – 100) range.

The second difficulty is that people express their perceptions differently on a scale. For example, many people do not use the scale edges, avoiding extreme positions. By doing so they narrow the actual range of the rating. These people do not use the scale in a linear way. Values in the middle are more widely used, so the middle values are more condensed, which makes it look as if the scale is not linear. On the other hand, some people always take extreme attitudes, avoiding the use of the middle scale. In general, using people as measuring instruments is really difficult, at least unless they can be "calibrated" (as is done in any case with instruments). To partially overcome these difficulties, reference samples can be given at the tasting, and in many cases they are anchored to certain positions on the scale. For example, in a very simple sour test (to compare the sourness of a list of samples), the very low and the very high concentrations of sour samples are presented, and are referred to as "1" and "10" on the scale. By doing so, two anchors are set on the scale. One anchor reference is also possible, for example in the middle of the scale (numeric or graphic). Using anchor references helps to homogenize the panel responses on a common basis, and to set a more reliable scaling spread.

In wine tasting this is, of course, impossible to do. The size of the scale is also important. A scale field of three (low, medium, and high, or bad, fair, and good) is very easy and convenient to use, but on the other hand, it is not sensitive enough to express a higher resolution of the sensations that people feel and are able to express. A field of ten would be much better, but for many people, as was said before, it is too wide. Experience shows that a field of 5 to 7 rating values in the scaling rate would be the most suitable, assuming the judges would really use the entire scale.

A rating test for evaluating one sensory dimension is relatively simple after training the panel to do the specific task of the rating—for example, to rate the relative sourness, sweetness, bitterness, saltiness, and astringency of a solution. All these sensations are compared by their intensity. The results also can be absolutely verified by their concentration in the solution. But how can we evaluate a sample like wine, which is completely different in two aspects: One, there are many parameters to evaluate within the same sample. These parameters are mixed together and interact with one another. The perception of a mixture of parameters is much more complicated than that of individual ones. Secondly, and very importantly in evaluating wine, most of the parameters are not quantitative ones but qualitative. The question here is, "How *good* is the smell of the sample?" instead of, for example, "How sweet is it?" Also, the taster has to find out the unique characters in the sample, instead of relating it to a given parameter. The evaluation has to rate the wines as a "whole" and to compare the wine samples by rating them in the order of "how good" they are. In addition, the sample should not be rated only within the narrow boundaries of the given list of tested samples, *but should also be related to the whole domain of similar wines in the world.* All these points are presented here (in a very short notes), in order to make clear that evaluating wine is definitely not simple and not as accurate as might be expected.

Also, we did not mention here the standard conditions of the tasting which can completely alter the results (e.g., glasses, temperature, light, blind tasting procedure, number of wine samples at one tasting session, score tasting procedure, result evaluation procedure and more).

Facing so many difficulties, some attempts have been made to facilitate wine evaluation, which we shall mention here. We shall not deal with technical procedures (which are very important), but rather with the heart of the question namely, the rating score. The principle in the rating score is to break down the major organoleptic responses (parameters) and evaluate them separately for their contributions to the wine under study by giving them a numeric score. Each parameter could have a maximum score according to its assumed importance (weight) in wine quality. By summing the individual scores, the total wine quality score can be obtained. There are many score tables offered in the literature. One of the most recognized is the *Davis Scorecard*, which was developed at the University of California in Davis. The original and the modified Davis scorecards are shown in the following table:

Original Score		Modified Score	
parameter	*weight*	*parameter*	*weight*
appearance	2	appearance	2
color	2	color	2
aroma, bouquet	4	aroma, bouquet	6
volatile acidity	2	astringency	1
total acidity	2	total acidity	2
sweetness	1	sweetness	1
body	1	body	1
flavor	2	flavor	2
bitterness	2	bitterness	1
general quality	2	general quality	2
total	**20**	**total**	**20**

Rating: 20 - 17 superior
16 - 13 standard
2 - 9   below standard
8 - 1   unacceptable

As the wine industry rapidly changed in California during 1950 – 1970, the original card was changed to the modified scorecard, which reflected the improvements in wine quality at that period. Although the Davis scorecard was widely used for many years, it has some serious weak points which limited its full use:

(1) *Too high* a weight was given to the look of the wine – *appearance* and *color* (20% of the total quality value).

(2) Too low a weight was given to the most important parameters, namely, *aroma*, *bouquet* and *flavor* (only 30% in the original scorecard, and 40% in the modified one).

(3) It is not clear how "*sweetness*" should be scored. Is it a "good" parameter which contributes to the wine quality, or not ?

(4) The absence of "*volatile acidity*" in the original card gave a wine 10% to the quality score, which looks odd in modern winemaking. It was omitted in the modified card.

(5) The same argument applies to "*bitterness*," which is not justified in today's technology and was not omitted in the modified card.

(6) The term "*astringency*," like sweetness, does not specify whether it is "good" or "bad," and how to relate to it.

(7) All of the above insignificant parameters, which by their absence in wine gave indications of quality (*sweetness, astringency, bitterness*), and which sum up to 15% in the modified card. These are credited to the wine quality just because of "good behavior."

(8) *Too low* a weight for "*general quality*," which includes the overall impression of the wine as an integral unity such as the balance, aging, after taste, off-flavor or other pros and cons.

(9) Also very important is the limited range of rating for each parameter (between 1 or 2 points, or at most 1/2), which does not leave enough space to express the full potential differences in each parameter. (Of course other fractions can be used, e.g., 1/3, 3/7, etc., but then many people might have some difficulties summarizing them.)

After 14 years of use, the Davis score card was re-evaluated by expert panel members for the purpose of modifying it. In practice it was shown that in most cases the numbers on the score span (12 – 16) points, as if the whole scale were shrunk into this range. It does not make wine evaluation easy if the rating falls into such a narrow range. Other attempts were also made to develop wine scoring cards, with different approaches to the scoring problems.

Facing the difficulties in forming any wine tasting card, we challenged the Davis card by proposing a new card which would include the following principles:

(**1**) The appearance (*color* and *clarity*) would have maximum of 10% of the total score.

(**2**) Aroma, bouquet and flavor would have 50% of the total score.

(**3**) A new parameter called "defect" is included, which will contain any major or minor defect that exists in the wine. It includes, for example, those "insignificant" parameters mentioned above, and any attribute which reduces wine quality such as mold, corkiness, over-oxidation, excessive oakiness, or astringency when it is very disturbing, bad aftertaste, mousy and any other defect which was found on tasting. The evaluation of this "*defect parameter*" is different than the others and will be discussed later.

(**4**) A "*general quality*" parameter (weight 20%), which will allow the judge to express his/her personal opinion on the wine quality, such as the wine's balance, mouth feel (body), aging and aftertaste.

(**5**) The *general quality* and *defects* parameters together amount to 30% of the total. It gives the judge a wide range of freedom to appreciate a good wine or to depreciate it when it gets too many quality points on less important parameters (which happens in many cases in simple well-made wines).

(**6**) The range of scoring *each* parameter was expanded to span from **0 ==> 5**. The taster deals only in a field of six (from 0 to 5) which gives enough resolution without going into 1/2 or 1/3 of a point, and it is not too large to get lost in too wide a range. Each parameter has its own *weight value* according to the principles listed above, namely, its relative importance in the wine quality rating. Therefore, the rating of each parameter is multiplied by its weight, and the total is summarized.

The total weight of the parameters is 20, and because the maximum rating of each individual parameter can be from 0 to 5, the maximum wine score might sum to **20 x 5 = 100 points**. The card was presented in the *Concepts in Wine Chemistry* book, and is shown here in the following table:

Name	Sample No.						
	Wine						
Note	Winery						
	Vintage						
Date	Price						
Appearance & Color (x2) brilliant / clear / turbid deep / regular / light							
Aroma & Bouquet (x6)							
Flavor (x4)							
Acidity (x2) high / medium / low							
Defects (*minus score*) (x2) acetic / oxidized / moldy oaky / corky / sulfide / bad aftertaste / bitter earthy / hard /							
General Quality (x4) body / balance / aged harmony / aftertaste							
Total							

Score each parameter 0 ===> 5 ; then multiply by weight factor

Rating:
100 - 96 Superior
95 - 92 Excellent
91 - 88 Good
87 - 84 Standard
83 -75 Below standard
74 - 0 Unacceptable

To use the card:

1. Each parameter is rated between 0 ==> 5 and multiplied by its weight factor.

2. The *defects* parameter is rated a full 5 points if *no defect* is found in the wine. But if one or more defects are present, the wine is degraded in rating according to the seriousness of the defect, down to zero rating. For example, if there is a minor bitterness, which is a slight flaw, it might be rated 4 in this case. If there is some very serious fault, it might be rated down to 1 or zero. This defects parameter is the "bad behavior" parameter, which deducts points only from wines with some problematic flaws.

3. The use of the *general quality* parameter is to express the *general* and *integral* impression of the wine, and to emphasize any special issue the taster wants to address (for good or for bad) by means of the appropriate score. This parameter has high weight, and it is the tool to make the difference between mediocre wines and great ones.

4. After rating each parameter, the total sum makes it possible to compare the wines' scores.

5. For blind tasting, the wine's number is written on the top of each wine column.

6. The details of each wine tasted in the flight (winery, vintage, price) can be disclosed after tasting.

In this scorecard there are direct judgments related to quality (Aroma & Bouquet, Flavor, and General Quality) which add up to 14/20 points (70%) of the maximum score. The remaining 30% (Appearance, Acidity, Defects) is related to "good or bad behavior." We found that this is the right balance between the main parameters in wine evaluation.

After a few years of using this card, we found that it was not fulfilling some of our expectations:

(1) The look of the wine has very much less importance than it has been given in the card (10% of the score). The brilliancy of white wine has nothing to do with its quality, unless a real defect has occurred in the bottle. This defect will show up anyway in the wine's smell and in the mouth, and will be evaluated there. The technology nowadays in producing clear white wine is very common, and it has nothing to do with wine quality. In red wines the clarity is difficult to see because of the red color, especially in dark heavy wines. Thus, it is irrelevant to the wine quality. The hue of the wine (white and red) reflects its age or its oxidation state. It is not a matter of quality. If the wine is over-oxidized, it will be very clearly shown in its bouquet and taste. If the wine is red-brown or white-yellow because of age, it will again be sensed by the nose and mouth. In both cases no real conclusion can be deducted from its appearance. In red wine production, it is quite common now not to filter the wine before bottling. How do we relate to it in the scoring card? And sediments in aged red wine bottles are normal and do not show any connection to its quality. We therefore find that these parameters of appearance and color, are irrelevant and should be eliminated.

(2) The acidity parameter is really not a quality term in itself. In a one-dimensional test it can only be judged by its concentration or intensity. Its real quality contribution to wine is in its balance with other parameters, such as sweetness, astringency and body. There is no justification to take it as quality parameter; instead, we should relate to it as part of the balance in the tested wine.

(3) As a matter of fact, we do not think that one should relate to *any* specific attribute such as acidity, bitterness, sweetness, astringency, body, oakiness, etc. as a quality parameter. None of these attributes stands by itself as a quality parameter. Only their combination in the evaluated wine determines its quality. Therefore, we suggest focusing instead on the *harmony* and *pleasant sensation* we feel by our nose and by mouth. The various parameters can be judged only by their contributions to this harmony for good or for bad. By doing so, we abandon the old method of differentiating wine quality by means of individual factors, and evaluate wine quality as an integral sensation stimulator.

(4) There is no relation to the specific *characteristics* (descriptive) of the tested wines in any scorecard. This part of evaluation is of most importance and should be included in the scorecard.

Based on these arguments, we suggest a new scorecard containing only four gross parameters. Three of them are positively perceived and the fourth is a negative one, because unfortunately, negative attributes do exist in wine. The parameters are: *Nose, Mouth-feel, Harmony,* and *Negative Attributes*. Each of them contains, of course, the familiar factors which compose wine quality:

Nose: The smell sensation which includes the varietal aroma and bouquet. Its weight is **x7**.

Mouth-feel: The "taste" sensation in the mouth which contains the varietal flavor, bouquet, body, and acidity. Its weight is also **x7**, equal to the *Nose*.

Harmony: The integral impression of the evaluated wine, which includes its appearance, balance, complexity, aging, aftertaste, and uniqueness. Its weight is **x6**.

Negative Attributes: The negative parameters which include any of the factors that reduce wine quality, such as volatile-acidity, corkiness, bitterness, overoxidation, excessive oakiness, repulsive off-flavor, opaque color etc. Its weight is (**-x7**) in order to emphasize the heavy negative impact that defects have on in wine quality due to defects.

The sum of the weights of the three positive parameters is **20**. Each one of them can be scored between zero to five, and hence the maximum score might be **20 x 5 = 100 points**. The negative parameter rate ranges from zero to five and therefore can reduce the score by (-7 x 5 = -35) points, which are *deducted* from the total score.

In this way, the ultimate wine with no defects could be rated 100 by summing up its three quality parameters. Defects are not counted unless we notice them, and then they reduce the score as needed to bring the wine quality evaluation to its actual rating. No "good behavior" exists in the rating, but "bad behavior" is punished if found. We feel that using these scoring principles to rate wines will better express the quality and technology of the contemporary wine industry.

The scoring chart also contains a column for describing the wine by using terms from the Wine Aroma Wheel. We believe this scorecard is very simple and clear to use. It emphasizes the virtues of wine quality and its characteristic descriptions, and will be helpful in the field of wine evaluation. The scorecard is shown in the following figure. Each card is for one wine sample.

Wine Tasting Evaluation

Date _____

Variety _____   *Vintage* _____   *Wine No.* _____

	Weight	Characteristics	Score 0 ==> 5	Value
Nose Varietal aroma Bouquet	x7			
Mouth feel Flavor Acidity	x7			
Harmony Color Balance Complexity Body Aging After taste	x6			
Negative Attributes Color / Acetic / Oxidized Moldy / Oaky / Bitter Corky / Repulsive aftertaste	-x7			

Total _____

Bibliography

General

1.   *Knowing and Making Wine*, E. Peynaud, J. Wiley, New York (1984).

2.   *Principles and Practices of Winemaking*, R. Boulton *et al.*, J. Wiley, New York (1995).

3.   *Concepts in Wine Chemistry*, Y. Margalit, The Wine Appreciation Guild, San Francisco, 3rd Ed. (2012).

4.   *The Microbiology of Wine and Vinifications (vol. I, II)*, P. Ribereau-Gayon, *et al.*, J. Wiley (2006).

6.   *Production Wine Analysis*, B. Zoecklein, *et al.*, Avi Books, New York (1994).

7.   *Wine Microbiology and Biotechnology*, G. H. Fleet, ed., Harvard Academic Publishers, Philadelphia (1993).

8.   *Wine Science,* R.S. Jackson, Academic Press, New York (1994).

9.   *Vine, Grapes and Wine*, J. Robinson, Alfred Knopf, New York (1986).

10. *Grape Growing*, R. J. Weaver, University of California Press, Berkeley (1976).

11. *Viticulture (vol. I, II)*, B.G. Coombe & P.R. Dry, eds., Winetitles, Adelaide (1992).

12. *General Viticulture*, A.J. Winkler, *et al.*, University of California Press, Berkeley (1974).

13. *Handbook of Enology (Vol. I, II)*, P. Ribereau-Gayon, *et al.*, J. Wiley, New York (2000).

14. *Terroir*, J. Wilson, University of California Press, Berkeley (1998).

15. *Sunlight into Wine,* R. Smart & M. Robinson, Winetitles, Adelaide (1991).

16. *The University Wine Course*, M. Baldy, The Wine Appreciation Guild, San Francisco (1997).

Wood and Cork

1.   Browning, B.L., *The Chemistry of Wood*, J. Wiley, New York (1963).

2.   Sjostrom, E., *Wood Chemistry, Fundamentals and Applications*, Academic Press (1993).

3.   Technological report, "Cork, naturally the most suitable stopper," *Aust. & New Zealand Ind. J.* 4, Feb 1989, p. 31.

4. Cooke, G.B., "Cork and its uses," *J. Chem. Educ.*, Aug 1931, p. 1462.

5. Cooke, G.B., *Cork and the Cork Tree*, Pergamon Press 1961).

6. Gibson L.J. et al., "The structure and mechanics of cork," *Proc. Royal Soc. London, A 377* (1981), p. 99.

Wine Evaluation

1. Amerine, M.A. and Roessler, E.B., *Wines, Their Sensory Evaluation*, W.H. Freeman, New York (1976).

2. Clapperton, J.F. et al., "Progress towards an international system of beer flavor terminology," *Master Brewing Assoc. of Am. Tech. Quart.* 12 (1975), p. 273.

3. Meilgaard, D.S. et al., "Reference standards for beer flavor terminology system," *J. Am. Soc. Brew. Chem. Proc. 40* (1982), p. 119.

4. Shortreed, G.W. et al., "The flavor terminology of Scotch whisky," *Brewer Guardian*, 108 (11) (1979), p. 55.

5. Piggott, J.R. and Jardine, S.P., "Descriptive sensory analysis of whisky flavor," *J. Inst. Brew. 85,* March 1979, p. 82.

6. Noble, A.C. et al., "Progress towards a standardizing system of wine aroma terminology," *Am. J. Enol. Vit. 35* (1984), p. 107.

7. Noble, A.C. et al., "Modification of a standardized system of wine aroma terminology," *Am. J. Enol. Vit. 38* (1987), p. 143.

8. Ough, C.S. and Winton, W.A., "An evaluation of the Davis wine score card and individual expert panel members," *Am. J. Enol. Vit. 27* (1976), p. 136.

9. Rankine, B., "Roseworthy develops new wine score card," *The Australian Grapegrower & Winemaker*, Feb. 1986, p. 16.

10. A. Waterhouse & S. Ebeler, *Chemistry of Wine Flavor*, American Chemical Society, Symposium Series (1999).

12. Peynaud E., *The Taste of Wine*, The Wine Appreciation Guild, San Francisco (1987).

INDEX

water additions, 251
U.S. legal minimum/maximum,
248–249
U.S. taxation and, 251
Alcohol fortification, 77
Alcohol(s)
fermentation by-products, 68–71
as fermentation inhibitor, 49, 50,
72, 77
yeast tolerance of, 56, 58, 220
See also Ethanol
Aldehydes, 71, 75, 144–145
Alkaline solutions, for cleaning, 123–
124
Allergenic effects, 220
Alsace bottles, 157–159
Aluminum haze, 228
American oak, 131, 143–144, 147–150
American Viticulture Appellation
(AVA), 120
Amino acids, 104, 108
Ammonia, 230
Ammonium carbonate, 252
Amyl alcohol, 69
Anaerobic spoilage, 237–241
Anthocyanins, anthocyanidins, 27, 30,
158, 183, 186–187, 190, 195,
197, 198, 203, 212
pH and, 191–192
reduced by gelatin fining, 109
SO₂ interaction, 224
tannin interactions, 33–34, 186–
187
Appellation
on label, 120
legal definitions, 249–250
Arabinose, 4, 74, 186
Arabitol, 70
Aroma
bentonite fining and, 105–106
blending to correct, 120–121
carbon fining and, 112

defined, 255
diacetyl and, 72
harvest timing and, 14
neutral esters and, 72, 73
from oak phenolics, 136, 143,
144–147
standardized aroma terminology,
260–261
See also Bouquet; Flavor; Nose
Aromatic aldehydes, 146
Aromatic ring, 180
Ascorbic acid, 253
Aspergillus, 17
Astringency, 27
defined, 255–256
fining and, 107, 109, 113
from oak phenolics, 144, 149
sources of, 27, 108
See also Tannins
ATP, 83

B

Bacteria
commercial freeze-dried, 86
microbial spoilage, 234–241
sulfur dioxide as antimicrobial,
215–218
See also specific types
Balling. *See* Brix
Barrel aging, 130–142
alternatives to, 151–153
French vs. American oak, 143–
144, 147–150
goals and effects of, 129, 143
over-aging, 234
racking during, 94–95
See also Oak phenolics
Barrel evaporation, 129, 136–138, 141
Barrel fermentation, 19–20, 51, 142,
147–150
Barrel maintenance, 138–142
cleaning, 123–124, 140–141

G

GAE (Gallic Acid Equivalent), 185, 194

Galacturonic-acid, 68

Gallic acid, 27, 143, 153, 180, 185
    oak astringency and, 144, 149
    polymeric forms, 185

Gallocatechin, 184, 189, 203

Gallotannins, 136, 188

Gelatin, 107, 108–109, 113, 254

General quality parameter, 266, 268

Generic wines, 120
    legal definition, 250

Geosmin, 168
    faults related to, 245

Geranium odor, 78, 238

Gewurztraminer, 74, 75, 106

Globalization, 212

Globulin, 110

Gluconobacter, 235–236
    See also Acetic acid bacteria

Glucose, 4, 74, 187

Glucosides, 186

Glycerol, 70, 74, 75, 79

Glycosides, 185–187
    See also Glucosides

Grape clusters, composition and yields, 16

Grape juice, preventing fermentation, 221–222

Grape leaves, 18, 190, 232

Grape ripening, 3–16
    acids, 7–11
    chemical changes, 3–4
    maturity determination, 12–14
    pre-harvest vineyard
        management, 17–19
    problematic, 13
    sampling, 12, 14–15
    stages in, 3
    sugar content measurement, 4–7
    weather conditions and, 12

Grape skins. *See* Skin contact

Grape stems, 25, 26, 40, 190

Grape sugar additions, 47, 74, 252

Grape wine, legal definition, 248

GRAS, defined, 252

Gravitational settling, 28

Guaiacol, 146–147, 151, 166–167, 168
    faults related to, 247

Gypsum, 45, 253

H

Hansenula, 237

Harmony, in wine rating tests, 269, 270

Harvest operations, 23–51
    destemming and crushing, 25–26
    free run and press run, 36–41
    grape sampling, 12, 14–15
    manual vs. mechanical
        harvesting, 18–19
    must corrections, 41–48
    skin contact, 27–36
    See also Pre-harvest operations

Haze. *See* Metal haze; Phenolic haze;
    Protein haze

Headspace, in bottle, 170, 175

Health warnings, on labels, 173, 251

Heartwood, 131

Heat exchangers, 26, 81

Heat of fermentation, 79–82, 197

Heat release, during fermentation, 49–50

Heat treatments, thermo-vinification, 33, 35

Hexanoic acid, 65, 86

Hexanol, 69

High alcohols (fusel oils), 68–71

Higher polymer phenols, 110

Horizontal presses, 37–38

Hot water, 122–123

Hot-water pressure machines, 21, 122–123

H_2S. *See* Hydrogen sulfide

61, 81
fining, 105, 109, 110, 111
glycerol content, 70
inoculation and fermentation
    procedures, 61, 63–64
methanol content, 68
oxidation, 233
polymeric phenolic content, 185,
    190
production yields, 16
recommended must acidity range,
    43
skin contact, 26, 27, 31–33
storage temperature, 125
TA measurement, 42
tank space needed during
    fermentation, 19–20, 61
tannins in, 108
time required for fermentation, 51
total phenolic content, 185
typical pH levels, 9
Reduction, of amino acids, 69
Redwood, 130
Refractometer, 7
Relative humidity, barrel evaporation
    and, 137–138
Residual sugar, 55, 73–78
    blending to correct, 121
    preventing fermentation after
        bottling, 77–78, 238
    sweetening dry wines, 77
    techniques for stopping
        fermentation, 76–77
    types of sweet wines, 74–76
Rhamnose, 4, 74, 186
Riesling, 74, 222, 223
Ripening. See Grape ripening
Rosé wines, 32
    fermentation temperatures, 50
    glycerol content, 70
    oxidation, 233
    total phenolic content, 185

See also Blush wines
Rotor-fermentors, 34, 61, 197–198
Rotten egg smell. See Hydrogen
    sulfide

S

Saccharomyces bailii, 222, 241
Saccharomyces bayanus, 56, 57, 58,
    220
Saccharomyces beticus, 237
Saccharomyces cerevisiae, 56, 57, 58,
    70, 220–221
Saccharomyces ellipsideus. See
    Saccharomyces cerevisiae
Saccharomyces fermentati, 237
Safety caution, during fermentation, 62
Salting-out, 98
Sampling grapes, 12, 14–15
Sanitation, 20–21, 122–125
    barrel cleaning, 123–124,
        140–141
Saturation, of air-oxygen in water, 199
Sauternes, 68, 75–76
Sauvignon Blanc, 39, 75, 94, 104, 106,
    120, 142, 190
Scorecards. See Rating tests
Sediments. See Lees
Semi-sweet wines, 74
Semillon, 39, 75, 120, 190
Settling tanks, 26
Shaving barrels, 151
Sherry, 74, 77, 147
    acetaldehyde content, 71
    alcohol content, 249
    flor yeasts, 71, 237
Sherry-like odor, 200
Silica gel (kieselsol), 107, 109, 113,
    254
Silicic acid, 107
Sinapaldehyde, 144, 145
Skin contact, 27–36, 198
    methanol content and, 68

WINE FAULTS
JOHN HUDELSON
$39.95 ISBN 978-1-934259-63-4

ACIDITY MANAGEMENT IN MUST & WINE
VOLKER SCHNEIDER
$ 978-1-935879-18-3

PRACTICAL FIELD GUIDE TO GRAPE GROWING & VINE PHYSIOLOGY
SCHUSTER, PAOLETTI, BERNINI
$ ISBN 978-1-935879-31-2

THE VITICULTURE & ENOLOGY LIBRARY

VIEW FROM THE VINEYARD
CLIFFORD P. OHMART
$34.95
ISBN 978-1935879909

UNDERSTANDING WINE TECHNOLOGY
DAVID BIRD
$44.95
ISBN 978-1-934259-60-3

CONCEPTS IN WINE TECHNOLOGY
YAIR MARGALIT
$45 ISBN 978-1-935879-80-0

CONCEPTS IN WINE CHEMISTRY
YAIR MARGALIT
$89.95 ISBN 978-1-935879-81-7

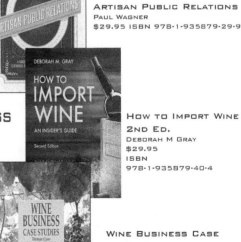

THE BUSINESS OF WINEMAKING
JEFFREY L LAMY
ISBN 978-1-935879-65-7

THE BUSINESS OF SUSTAINABLE WINE
SANDRA TAYLOR
$
ISBN 978-1-935879-30-5

WINE MARKETING AND SALES 3RD ED.
WAGNER, OLSEN, THACH
$75 ISBN 978-1-935879-44-2

THE WINE BUSINESS LIBRARY

ARTISAN PUBLIC RELATIONS
PAUL WAGNER
$29.95 ISBN 978-1-935879-29-9

HOW TO IMPORT WINE 2ND ED.
DEBORAH M GRAY
$29.95
ISBN 978-1-935879-40-4

WINE BUSINESS CASE STUDIES
PIERRE MORA, EDITOR
$35.00
ISBN 978-1-935879-71-8